Human Evolution

TERTIARY LEVEL BIOLOGY

A series covering selected areas of biology at advanced undergraduate level. While designed specifically for course options at this level within Universities and Polytechnics, the series will be of great value to specialists and research workers in other fields who require knowledge of the essentials of a subject.

Recent titles in the series:

Social Behaviour in Mammals	Poole
Genetics of Microbes (2nd edn.)	Bainbridge
Seabird Ecology	Furness and Monaghan
The Biochemistry of Energy Utilization in Plants	Dennis
The Behavioural Ecology of Ants	Sudd and Franks
Anaerobic Bacteria	Holland, Knapp and Shoesmith
Biology of Fishes	Bone and Marshall
The Lichen-Forming Fungi	Hawksworth and Hill
Seabird Ecology	Furness and Monaghan
An Introduction to Marine Science (2nd edn.)	Meadows and Campbell
Seed Dormancy and Germination	Bradbeer
Plant Growth Regulators	Roberts and Hooley
Plant Molecular Biology (2nd edn.)	Grierson and Covey
Polar Ecology	Stonehouse
The Estuarine Ecosystem (2nd edn.)	McLusky
Soil Biology	Wood
Photosynthesis	Gregory
The Cytoskeleton and Cell Motility	Preston, King and Hyams
Waterfowl Ecology	Owen and Black
Biology of Fresh Waters (2nd edn.)	Maitland
Tropical Rain Forest Ecology (2nd edn.)	Mabberly

TERTIARY LEVEL BIOLOGY

Human Evolution

ALAN BILSBOROUGH

Professor of Anthropology
University of Durham

BLACKIE ACADEMIC & PROFESSIONAL
An Imprint of Chapman & Hall
London · Glasgow · New York · Tokyo · Melbourne · Madras

Published by
Blackie Academic & Professional, an imprint of Chapman & Hall,
Wester Cleddens Road, Bishopbriggs, Glasgow G64 2NZ, UK

Chapman & Hall, 2–6 Boundary Row, London SE1 8HN, UK

Blackie Academic & Professional, Wester Cleddens Road, Bishopbriggs,
Glasgow G64 2NZ, UK

Chapman & Hall, 29 West 35th Street, New York NY10001, USA

Chapman & Hall Japan, Thomson Publishing Japan, Hirakawacho Nemoto
Building, 6F, 1-7-11 Hirakawa-cho, Chiyoda-ku, Tokyo 102, Japan

DA Book (Aust.) Pty Ltd., 648 Whitehorse Road, Mitcham 3132, Victoria,
Australia

Chapman & Hall India, R. Seshadri, 32 Second Main Road, CIT East,
Madras 600 035, India

First edition 1992

© Chapman & Hall, 1992

Typeset in 10/12 pt Times New Roman by Thomson Press (India) Ltd.,
New Delhi
Printed in Great Britain by T.J. Press (Padstow) Ltd.

ISBN 0 751 40077 7 (HB) 0 751 40078 5 (PB)
      0 412 02111 0 (USA) (HB) 0 412 02121 8 (PB)

A catalogue record for this book is available from the British Library

Library of Congress Cataloguing-in-Publication data available

∞ Printed on permanent acid-free text paper, manufactured in accordance
with the proposed ANSI/NISO Z 39.48-199X and ANSI Z 39.48-1984

Preface

Human origins exert a perennial fascination which now, due to media attention, touches more people than ever before. This account attempts a rather different perspective from many others in giving a fairly full survey of the actual evidence for human evolution with sufficient detail to indicate the major hominid groupings, the contrasts between them and, equally important, their internal variability. In addition, I have tried to distinguish description from interpretation or at least sign-post them sufficiently for readers to recognise the different issues involved in these complementary activities. Evolutionary processes generate complex phenomena that are only partly represented by the available evidence, and we all have theoretical preconceptions that influence how we view the record. This means that the same evidence can quite legitimately be interpreted in contrasting ways by different workers. Some find this situation confusing and 'unscientific' but for many it is a prime factor in the subject's fascination. I have therefore tried to summarise the range of models currently under debate while also making clear where my own preferences lie. In this way readers will, I hope, have sufficient information to make their own evaluations and identify both the strengths and weak spots of particular interpretations.

The book is planned as a relatively self-contained account which includes brief summaries of dating techniques, evolutionary and phyletic principles, and aspects of functional anatomy as background information. However, the main emphasis is on the hominid fossil record of the last 5 million years and its interpretation, with the bulk of the text given over to this. There are also summaries of the contextual evidence necessary to build up a picture of hominid environments and adaptations; as part of that picture insights into hominid behaviour obtained via the archaeological record are also included. In an attempt to escape from the constraints imposed by taxonomic thinking I have deliberately ordered the material in broad chronological bands rather than by taxa. This means that some groups (e.g. the robust australopithecines) are treated in several chapters, but has the advantage that they can then be viewed, along with other synchronic forms, as part of a broader hominid array, and their interactions stressed. In the final chapter I have attempted to draw the threads together, identify some expanding areas of enquiry, and venture to predict likely developments over the next few years.

I am grateful to many people for their assistance in the book's preparation. The work of many colleagues and friends has provided stimulus and challenge over the years, but I should particularly like to acknowledge the contributions of Leslie Aeillo, Peter Andrews, Michael Day, Rob Foley, Jim Garlick, Geoff Harrison, Richard Leakey, the late Charles McBurney, Chris Stringer, Phillip Tobias and Bernard Wood. I owe a particular debt to Bernard Wood for allowing me to incorporate a summary of his findings on the Koobi Fora hominids prior to their publication. Of course, none of the above can be held responsible for what follows, which they might well want to disown. I also thank my students past and present for the insights provided by their comments and questions, particularly the ones I couldn't answer.

I am grateful to Trevor Woods and Bob Read for photographs, and to David Jobson and students at Sunderland Polytechnic who prepared the line drawings: Janet Adams, Linda Amos, Carolyn Bellas, Jackie Farrand, Michael Hudson, David Mitchell, Andrew Pearson, Antony Stobbart, Sarah Walker, Katie West and Fiona White. It is customary for authors to proffer thanks for secretarial help; only those who have laboured dispiritingly to decipher my hieroglyphics know how merited such acknowledgement is in this case. I owe a major debt to Sue May and Muriel Borradaile for secretarial valour above and beyond the call of duty.

Finally, and most of all, I thank Lynn, Louise and Fiona for putting up with the book's production, and with me for much longer.

A.B.

Contents

Chapter 1   INTRODUCTION: THE SCOPE OF
             PALAEOANTHROPOLOGY   1

Chapter 2   EVOLUTIONARY AND PHYLETIC
             PERSPECTIVES   5
             2.1   Introduction   5
             2.2   Evolutionary mechanisms   6
             2.3   Evolutionary models   8
             2.4   Phyletic principles   10
             2.5   Classification   18
             2.6   The temporal framework   21
                    2.6.1   Carbon-14 (radiocarbon)   22
                    2.6.2   Potassium–argon (K–Ar) and ^{40}Ar–^{39}Ar 23
                    2.6.3   Thermoluminescent (TL) and electron spin
                             resonance (ESR) dating   24
                    2.6.4   Amino acid racemisation dating   25
                    2.6.5   Uranium series dating   25
                    2.6.6   Palaeomagnetism   25

Chapter 3   HOMINID STRUCTURE AND
             FUNCTION   28
             3.1   Introduction   28
             3.2   Biomechanics   28
             3.3   Head and skull   29
             3.4   Dentition   31
             3.5   Masticatory activity   34
             3.6   Transmission of chewing forces and mid-face structure   36
             3.7   Vocalisation   37
             3.8   Brain   38
             3.9   Trunk   44
             3.10   Pelvis and hindlimb   46
             3.11   Bipedalism   49
             3.12   Forelimb   50

Chapter 4   THE CATARRHINE RADIATION AND
             HOMINID ORIGINS   54

Chapter 5   PLIOCENE HOMINIDS   68

    5.1   Introduction   68
    5.2   Historical   68
    5.3   Australopithecine species   71
        5.3.1   *A. africanus*   73
        5.3.2   *A. robustus*   75
        5.3.3   *A. boisei*   78
        5.3.4   *A. afarensis*   81
        5.3.5   Omo   84
        5.3.6   *A. aethiopicus*   87
    5.4   Cranio–facial biomechanics   87
    5.5   Postcranial material   91
    5.6   Growth, development and body size   97
    5.7   Evolutionary relationships   99
    5.8   Australopithecine environments and adaptation   103

Chapter 6   PLIO-PLEISTOCENE HOMINIDS   109

    6.1   Introduction   109
    6.2   East Africa   110
        6.2.1   Olduvai Gorge   110
        6.2.2   Lake Turkana   114
        6.2.3   Omo   121
    6.3   South Africa: Transvaal sites   121
    6.4   Systematic interpretations   123
        6.4.1   *Homo habilis*   123
        6.4.2   Diversity in early *Homo*   124
    6.5   *Australopithecus* or *Homo*?   130
    6.6   Phylogeny   131
    6.7   Summary   134
    6.8   Robust australopithecines   135
    6.9   Environments and habitats   135
    6.10 Behaviour   136

Chapter 7   LOWER AND MIDDLE PLEISTOCENE
           HOMINIDS   145

    7.1   Introduction   145
    7.2   Early *Homo erectus*   149
        7.2.1   Asia   153
        7.2.2   Stability or change in *H. erectus*   162
    7.3   Archaic *Homo sapiens*   165
        7.3.1   Sub-Saharan Africa   167
        7.3.2   North Africa   169
        7.3.3   Europe   170
        7.3.4   Asia   172
    7.4   *H. erectus* or *H. sapiens*?   173
    7.5   Ecology and behaviour of Lower and Middle Pleistocene
        hominids   175

Chapter 8   UPPER PLEISTOCENE HUMAN
             EVOLUTION   182
             8.1   Introduction   182
             8.2   Chronology and environment   184
             8.3   Neanderthals and the archaic modern transition   186
                    8.3.1   Western Europe   195
                    8.3.2   Central and eastern Europe and Russia   198
                    8.3.3   Middle East   200
                    8.3.4   Far East   204
                    8.3.5   Africa   209
                    8.3.6   The Americas   212
             8.4   Contemporary evidence   213
             8.5   Interpretations   215
             8.6   Selection   219
             8.7   *Homo sapiens sapiens*   222

Chapter 9   HUMAN EVOLUTION: PATTERNS,
             PROBLEMS AND PROSPECTS   224

REFERENCES   235

INDEX   251

CHAPTER ONE
INTRODUCTION: THE SCOPE OF PALAEOANTHROPOLOGY

Palaeoanthropology, the study of human evolution, is currently in a dynamic, expansionary phase, and the subject bears little resemblance to that of only 20 years ago. This development arises from a number of causes. Fossil discoveries have burgeoned in recent years so that there is now much more hard evidence for human evolution compared with even two decades ago. This greatly extended fossil record has refined interpretations of later human evolution, and prompted radical, almost revolutionary, re-appraisals of the earliest phases. However, these advances have not followed just from accumulating fossil evidence, for a denser fossil record would be of little value without new methods of analysis and interpretation. The revolution in palaeoanthropology follows as much from analytical developments and conceptual advances as from fossil discoveries.

One characteristic of current studies is their collaborative, multidisciplinary nature. While the palaeoanthropologist has traditionally been viewed as a lone figure in the field or laboratory, current investigations involve specialists from a variety of disciplines within anthropology, archaeology, the biological and medical sciences, chemistry, physics and the earth sciences. Nor are these workers simply providing new evidence within the traditional framework: the scope of enquiry has broadened enormously. Whereas earlier workers were largely content to describe the fossil evidence and leave it at that, current studies are much more concerned with putting flesh on the bones and setting them in context, so as to reconstruct earlier hominids existing in particular past habitats, following particular adaptive strategies and responding to distinctive selection pressures.

In order to reconstruct this broader picture, the contextual association of fossils, yielding information about their age and the related landscape, climate, animal and plant communities, becomes as important as the specimens themselves. Some examples, by no means exhaustive or comprehensive, are given to illustrate the range of methods and areas of enquiry that characterise current palaeoanthropology.

The study of fossil specimens has been aided by new techniques in addition to the traditional methods of anatomy and morphometry. These new techniques include scanning electron microscopy (SEM), computerised axial tomography (CAT) scans, moiré fringe analysis, fourier transforms and other computer-aided techniques for biomechanical modelling (Oxnard, 1984), all of which have provided valuable data on the microstructure of tooth and bone, hominid growth and development, movement, mastication and diet. Chemical analysis can distinguish postmortem changes due to fossilisation from those reflecting dietary and pathological states during life, so yielding further insights into the life patterns of early hominids.

Nor are fossils the only source of information: a large (and growing) database of size–shape relationships in modern primate species helps to distinguish changes in fossil structures that are driven by contrasts in body size from those that are independent of overall size effects. There are important ecological corrolates which influence, for example, dietary strategies, and, increasingly, provide evidence of strong associations with life history patterns and behaviour (Smith, 1989). In addition, biochemical and molecular data have provided new information about the interrelationships of living hominoid species and the likely times of divergence of their lineages from a common ancestor, so enabling the fossils to be set within a broader phyletic framework. The calibration of hominoid evolution has been further aided by developments in physics that have led to new dating methods, while techniques in geophysics and geochemistry make for finer and more reliable cross-correlation between sites. In East Africa geochemical 'fingerprinting' of volcanic ash (tuffs) has resolved numerous dating uncertainties and aided site correlation, whilst in South Africa and Europe research into the processes of cave formation has clarified the contexts of numerous hominid specimens.

Another area of major development has been taphonomy, the study of factors affecting the integrity, persistence and preservation of carcasses after death, which aids identification of hominid and non-hominid agencies affecting, say, animal remains as potential food sources. Increasing awareness of the taphonomic and geophysical factors influencing site formation has in turn led archaeologists to define more carefully the criteria for establishing hominid tool-making activity from archaeological evidence.

Finally, there is growing recognition that to study fossil hominids alone is insufficient because early hominids did not exist in isolation but formed part of a wider ecological community with complex interactive effects.

Greater knowledge of the evolution of other animal groups—suids, bovids, elephants, other primates—will aid definition of hominid evolutionary biology, and a community ecology approach to early hominids can yield valuable insights. Once the ecological parameters are defined, models derived from comparative socio-ecological studies of modern primates will help reconstruct possible patterns of early hominid social organisation and behaviour.

Concurrent with, and in part because of, these developments, there has been increasing recognition that virtually any aspect of palaeoanthropology involves complex judgements based on assumptions and theoretical viewpoints that have traditionally gone unstated. There is now greater awareness of the need for a tighter, more explicit conceptual framework than in the past, so that workers' reasoning, as well as their evidence, can be evaluated. This can perhaps be most clearly seen in the debate over methods of phyletic inference and perceived patterns in the fossil record (chapter 2), but it extends to virtually all areas of the subject. For example, accounts of human evolution, including this one, usually interweave reconstructions of early hominid adaptation and behaviour with their schemes of evolutionary relationships, but it is important to recognise that several different and distinct areas of enquiry are involved here (Tattersall and Eldredge, 1977). These include: pattern recognition in the fossil record and the grouping of specimens; constructing evolutionary schemes from those groups; and relating the groups to environmental and other data to construct adaptive 'scenarios' that set the forms in ecological context. Only by distinguishing between these different procedures can the strengths and weaknesses of both data and argument be recognised, and a more rigorous treatment provided. Similar clarification is apparent in taphonomic and archaeological studies, with the aim of formulating testable hypotheses rather than narrative 'just-so' stories (Binford, 1981).

As an attempt to reflect the rapidly changing, varied nature of palaeoanthropology, this book has several aims. One is to present an overview of current knowledge of human evolution with an emphasis on the fossil record—the hard data of the subject, and the basis of all else. Another is to outline the varied conceptual frameworks within which the fossils are interpreted, and to show how different theoretical perspectives influence the judgements reached. The consequence is that evolutionary schemes are not absolute truths but models that aim to be maximally compatible with existing data and reflect particular aspects of phylogeny, but which are subject to modification or rejection as more data accumulate.

Another aim has been to convey some idea of within-group variability as well as the inter-group differences which are usually emphasised. Groups are not self-evident in the fossil record; it is the diversity of individual specimens that provides the basis for recognising groupings, but once these are established and formally named it is all too easy to then consider them as fixed and internally invariant. Knowledge of intra-group variation is, however, essential for assessing the integrity of claimed groupings, for investigating the significance of patterns of individual differences and possible sexual dimorphism, and for analysing aspects of evolution and adaptation.

This account also attempts a summary outline of hominid palaeoecology and behaviour, stressing that the fossils' morphologies and contexts point in many cases to niches and adaptations quite distinct from those of present-day species. Fossil hominids should be viewed as forms in their own right, responding to the demands of their particular habitats, rather than as incomplete and inadequate precursors of later forms (ourselves) that are possibly, and only possibly, their descendants.

Given these aims, this account dwells on gaps, controversies and uncertainties as much as it does upon the apparent certainties of the subject. It is hoped that readers will not find it confusing as a result; an impression of greater certainty would have meant a withdrawal from reality. By the time the book appears some of the gaps may well have been filled and controversies resolved as a result of new discoveries, but it is equally likely that some apparently solid foundations will have been undermined by those same discoveries: that is the nature of palaeoanthropology. If what follows conveys some of the interest and excitement of working in such a fast-moving, developing and expanding subject, it will have achieved its primary purpose.

CHAPTER TWO

EVOLUTIONARY AND PHYLETIC PERSPECTIVES

2.1 Introduction

Humans are extraordinarily adaptable and wide ranging, displaying great flexibility of response to ecological and environmental pressures in varied habitats. Our success is founded upon a particular set of features, some shared with other species, especially other primates, although modified in distinctive ways, and others that are unique. Primate features that have been critical for human success include: visual acuity (overlapping visual fields and hence stereoscopic vision, and sensitivity to light of different wavelengths, i.e. colour); dextrous extremities with mobile digits, nails instead of claws, and improved tactile sensitivity; brain enlargement and good eye–hand–brain coordination; a generalised dentition, efficient masticatory apparatus and a digestive tract capable of processing a range of items including insects, many plant products and vertebrate flesh; a reproductive strategy that involves reduced litter size (single births), efficient placentation, long intra-uterine and postnatal development periods that are associated with complex parental (maternal) care and elaborate social behaviour.

Many of these features can be viewed as adaptations to arboreal, small-branch niches in stable tropical environments, but they are clearly preadaptive to other niches, including terrestrial ones in more varied habitats, and so have permitted several other primates species (e.g. baboons, geladas, macaques and chimpanzees) as well as ourselves to effect the transition from tree to ground. Some of these features have been modified in humans to an extent not seen in other primates, for example, intra-uterine and early postnatal brain development, parental care and manual dexterity. In other features we contrast with most, if not all, other primates.

Any consideration of human adaptive success must take account of both similarities and differences to other primates, and emphasis on one at the expense of the other necessarily results in an incomplete analysis. However, reconstruction of evolutionary relationships and the search for earlier ancestors or relatives requires a shift of emphasis (see below). The important point to note here is that it is the contrasts which serve to distinguish humans from other primates and which, when present in fossils, indicate them to be hominid. The fossil record will reveal evidence of their evolutionary sequence, so that early hominids will possess one or more, but not necessarily all, of the characters that distinguish modern humans: it is these traits that identify the fossil as hominid, no matter how 'ape-like' its overall features. In this sense early evolutionists' concept of a 'missing link', ape-like in half its characteristics, human in the remainder, is hopelessly misconceived; just one human characteristic is sufficient to proclaim hominid status.

Distinctive human features include: true truncal erectness with associated changes of the shoulder and pelvic girdle; bipedalism so that the hindlimbs are exclusively involved in locomotion and the forelimbs freed for manipulation leading to a fully developed precision grip; a dentition and jaw adapted for efficient crushing and grinding of food items; throat anatomy that permits production of a wide range of vocal signals and hence the development of language; and a greatly expanded and complex central nervous system, and hence markedly more elaborate behavioural patterns. The functional basis for these features is outlined in the next chapter. They, and other human characters, represent the current outcome of evolutionary processes and shifting selection pressures on hominid and pre-hominid populations.

2.2 Evolutionary mechanisms

At the simplest microevolutionary level, evolution represents changes in the gene pool of a population over time: since the members of one generation do not contribute equally to the next, the composition of the population, and of its gene pool, will change, i.e. evolve. Such change will partly reflect random factors, especially if the population is small, but over a longer period by far the most potent determinant of evolution is natural selection. Since individuals differ some will possess features that better equip them to operate in particular habitats, the measure of 'better' being the number of offspring produced (reproductive success). Those who produce most offspring will obviously contribute maximally to the gene pool of the next generation and in this way the population evolves.

No environment is entirely constant for any length of time, but some (e.g. tropical ones) are more stable than others (e.g. those in high latitudes). Selection pressures in stable environments progressively 'fine tune' populations to their surroundings, so that most individuals are pretty well adapted, and the major component of selection will be stabilising selection, which refines and reinforces the status quo. A K-type reproductive strategy of small litter size, long development and pronounced parental investment is associated with stable environments. In fluctuating environments selection is often directional, favouring those individuals (perhaps a small minority) whose features best equip them to reproduce in the new, changed environmental circumstances. Markedly fluctuating environments elicit contrasting r-type reproductive strategies of large litter size and rapid maturation to maximise the genetic and phenotypic diversity of the population, so ensuring that some individuals are able to operate effectively in the new conditions.

Whereas evolution is a property of *populations*, selection operates on the phenotypic features of *individuals*. Suggested instances of *group selection* where disadvantageous effects on individuals are outweighed by favourable consequences for the group as a whole, notably in socio-ecological studies, especially the evolution of altruistic (selfless) behaviour, have foundered through lack of both convincing examples and adequate theoretical models to account for their occurrence. Altruism is more convincingly explained by *kin selection*, i.e. the reproductive advantage of such behaviour for relatives, who share a high proportion of the altruistic individual's genes.

Other individuals of its own and other species are components of the organism's environment. It is thus useful to view different species in the same habitat as part of an ecological community evolving together, since changes in one species will inevitably interact with and influence others, which in turn affect the first. Such *co-evolution* is most clearly seen in host–parasite and predator–prey relationships, but can be broadened to include the larger biotic community and its evolution, e.g. the success of the rapidly diversifying Pliocene monkeys placed constraints on the evolution of early hominids (see chapters 4 and 5). Community ecology is, in fact, proving to be an illuminating approach to interpreting early hominid diversity (Foley, 1984, 1987).

While microevolutionary examples can be studied in the field, larger scale processes cannot, and there is debate as to whether macroevolutionary phenomena of the kind revealed by the fossil record simply represent the cumulative effects of intergenerational change or the extent to which other factors are involved.

One such factor is speciation, the splitting of a single species into two or more daughter lineages reproductively isolated from each other. An established mechanism for speciation is geographical isolation (allopatric speciation) in which a subpopulation, perhaps occupying a peripheral part of the range, becomes isolated from the rest of the species by a geo-graphical/ecological barrier of some kind. Habitats on either side of the barrier differ so that the populations diverge as a result of contrasting selection pressures acting on different gene pools. This continues to a point when even if the barrier is removed and the populations come into contact again reproductive isolation has developed, i.e. speciation has occurred.

Instances of speciation without geographical isolation (sympatric speciation) have been proposed from time to time and possible genetic mechanisms, e.g. assortative mating, developed but none is wholly convincing. Parapatric speciation, involving semi-isolated populations, has also been proposed but needs to be more fully documented.

Multiple speciation events are evident in, for example, the radiations discernible in the fossil record. If speciation is common, and its direction random, then the factors influencing which species groups persist will be the speciation rates associated with particular lineages and the tendency for some species to survive for relatively long periods of time. The result is species selection (Stanley, 1979), which is an important aspect of the punctuational model of evolution (see below) and should not be confused with group selection.

2.3 Evolutionary models

Theoretical perspectives about the nature, pattern and tempo of evolution powerfully influence perceptions of the patterns discernible in the fossil record. The longest established view is that of 'phyletic gradualism', which in its essentially modern form is a product of the 1930s 'new synthesis' that integrated population genetics with Darwinian interpretations of organic diversity. Phyletic gradualism views evolutionary change as a product of both speciation events ('splitting') producing discrete lineages, and signifi-cant change within lineages without speciation (anagenesis or phyletic evolution) as a result of differential fitness. Immediately following speci-ation most lineages differ only slightly; subsequent divergence results from anagenetic change in each lineage. Since the fossil record is incomplete, anagenesis may result in arbitrary divisions of a lineage, recognised as chronospecies (see below); there may also be major gaps, perhaps indicated by the sudden appearance of a new group, representing an explosive phase

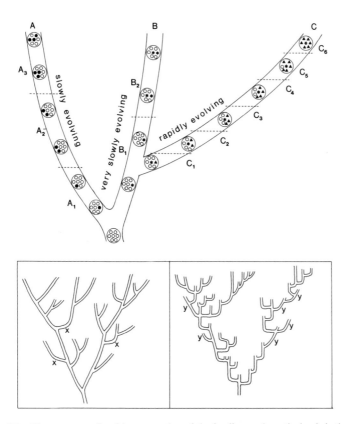

Figure 2.1 Upper: conventional interpretation of the fossil record: vertical axis is time and horizontal axis is overall divergence. Lineages A and B are derived from a common ancestor (open circles). A, identified by blacked circles, evolves relatively slowly, with chronospecies A_1–A_3 progressively more derived (more blacked circles). B (stars) evolves even more slowly, with only two chronospecies. Lineage C (triangles) evolves rapidly and so 6 chronospecies are recognised. The presence in C of traits characteristic of lineage B (1 star) shows C to be derived from B, not from the original common ancestor. Despite their marked divergence, B_2 and C_6 share common ancestry more recently than either does with A. See also Figure 2.2. Lower: alternative models of evolutionary change. Left: phyletic gradualism. Change results from both speciation and anagenesis, but with the latter predominating. While speciation sometimes produces marked differences (x), most species initially differ little from each other, and divergence occurs mainly through anagenesis. Right: punctuated equilibria. Most change is concentrated around speciation events which are much more frequent than on the previous model. Anagenesis sometimes occurs (y), but most lineages are stable until further speciation or extinction.

(adaptive radiation) involving both speciation and rapid anagenesis. If selection pressures are intense, numbers will be few and therefore poorly represented as fossils, hence the gaps.

This view has been challenged by the 'punctuated equilibrium' model (Eldredge and Gould, 1972) which views the fossil evidence as an unbiased record of what actually happened, i.e. as complete at any one time as at any other. The model posits that most evolution results from real speciation events, usually involving small, isolated, populations, perhaps close to the limits of the species range. Allopatric speciation is therefore seen as the primary 'engine' driving evolution. Significant change other than that associated with speciation does not usually occur so there is little anagenesis, most of its apparent instances within the fossil record reflecting only 'noise'—short-term fluctuations about a stable mean. Similarly, apparent long-term trends, often cited as supporting anagenesis, result from differential speciation rates and differential longevity of lineages ('species selection', see above). Evolution on this model is episodic, rapid change associated with speciation followed by periods of stasis. A corollary is that all species, including fossil ones, are real, since they are bounded by real biological events (i.e. speciation and further speciation or extinction), and the arbitrary division of a continuum involved in phyletic gradualism occurs infrequently. A further corollary is that lineages, speciation and extinction events are much more numerous than on the gradualist model.

Differences between the two models have sometimes been exaggerated and many workers consider elements of both models to be applicable. However, the punctuationists' view is that significant change is only associated with speciation, whereas the gradualists' view of anagenesis as a major determinant indicates fundamentally different interpretations of evolutionary mechanisms. Human evolution has been cited as a particularly clear and compelling example of both punctuated equilibrium (Gould and Eldredge, 1977; Eldredge and Tattersall, 1982) and phyletic gradualism (Cronin et al., 1981) which perhaps tells us a good deal about the nature and extent of the hominid fossil record (see also chapter 8).

2.4 Phyletic principles

Macroevolutionary patterns discerned in the fossil record will, of course, depend upon how that record is sorted and arranged. How best then to reconstruct past evolutionary events and relationships from the diversity of fossil and living species? The basic principles are straightforward enough, but there are alternative approaches to the detailed procedures of *phyletic inference* that have recently been much debated, and there is greater awareness of the need to explicate the aims, assumptions and procedures behind any given scheme: in large part the methods adopted depend upon the aims of the investigation and the emphasis given to different evolutionary phenomena.

The starting point for all phyletic reconstruction is the simple principle that in general forms closely resembling one another are closely related in evolutionary terms. In other words, their overall (phenetic) resemblance is a consequence of their close evolutionary (phyletic) affinity and so the former can be used as a rough and ready measure of the latter. If this was not the case we would never be able to get started, since in reconstructing human phylogeny (for example) we would have no reason to focus on fossil primates as the most likely candidates for ancestry, as opposed to fossil fish, spiders or even protozoa. However, there are obvious exceptions to the general equation that similarity equals phylogeny, and some sorting of the data on resemblance is required, the process of *phyletic weighting*. For example, homoplastic resemblance resulting from convergent or parallel evolution, e.g. the wing of birds and bats, the body form of whales and fishes, is phyletically misleading and needs to be omitted from assessments of evolutionary relationships; convergent resemblance is, in fact, much more widespread than often realised.

Similarly, characters that are functional or logical correlates of others provide no new information and should not be treated as if they do. The size and form of a lower canine tooth are specified by the upper canine since they together act as one functional unit and hindlimb length is made up of thigh and lower leg lengths, so that if these are included total hindlimb length should not be included as a further trait.

Even after excluding homoplastic and correlated characters, in any comparison many similarities are likely to be due to invariant primitive characters not directly relevant to the detailed phylogeny being investigated, e.g. similarities between humans, apes, monkeys and lemurs, such as bilateral symmetry, a vertebral column, homoeothermy, etc., do not help in unravelling evolutionary relationships within the order Primates. It is, in fact, the nested sets of 'advanced' or derived characters that successively identify together groups of organisms and eventually individual lineages. Once these are recognised, the presence of their identifying characters in new fossils allows these specimens to be assigned to particular lineages. A chronological framework will aid in distinguishing between lineages that are changing rapidly and those that are not, and will also determine the direction of change within lineages, so that information on the tempo of evolution and rates of diversification can be obtained.

The above summary describes the basic procedures of evolutionary systematics, or the 'stratophenetic' approach to phylogeny, which is concerned not only with identifying lineages and their members, but with reconstructing ancestor–descendant sequences. Such sequences may involve speciation and the production of new lineages (cladogenesis), or change within particular lineages (anagenesis) so that the later members differ from the earlier ones although no speciation has occurred. If anagenetic change is particularly marked, the fossils documenting the changes are likely to be assigned to different species because they differ so much, but species within a lineage ('vertical' species or chronospecies) are arbitrary and usually reflect nothing more than gaps in the fossil record. The effects of any gaps are obviously more marked in the record of rapidly evolving lineages, but in any event chronospecies should not be confused with 'real' (i.e. reproductively isolated) horizontal species (biospecies) corresponding to different lineages. In practice the great majority of living species have been recognised on morphological evidence rather than through breeding or genetic criteria; fossil species can, of course, only be identified this way. Such morphospecies at any given time (past or present) are generally assumed to equate with biospecies but there may be far from exact correspondence in particular cases.

Evolutionary rates differ between clades and also within lineages from time to time so that it can be dangerous to extrapolate from one group to another, or from documented phases of a lineage to where there are gaps in the fossil record. Rates also differ from structure to structure, leading to the phenomenon of mosaic evolution. For example, in human phylogeny truncal erectness and bipedal locomotion evolved well before brain expansion or facial reduction. Taxonomic boundaries (genera, species) usually do not correspond to major changes in all or even most systems, and may reflect nothing more than the fortuitous sequence of recovery of the fossil record, i.e. those chronospecies initially recognised.

Palaeoanthropology and primatology have suffered particularly from the tendency to view all extant and fossil forms solely in terms of their relevance to human ancestry. This has two unfortunate consequences. First, the search for ancestors, especially human ancestors, tends to predispose workers to place all fossils on the lineages leading to one or another of the extant forms. Where, as with hominoids (apes and humans), there are few living species, this not only leads to spurious ancestor–descendant sequences but can very seriously underestimate past diversity since it does not allow for past radiations in which the majority of lineages become extinct. In fact, there is good evidence that ape diversity in the Miocene (10–20 million years ago) was much more extensive than today, and that early hominid evolution involved several lineages, not one as at present (see chapters 4–6).

The second disadvantage of anthropocentrism is that it encourages a view of primate and hominid evolution as a directed, unfolding progression of increasing complexity with ourselves as the natural climax to that process. Apes (and early hominids) then tend to be viewed as incomplete humans, monkeys as incomplete apes, and so on. Such grades (groups with a common level of organisation or adaptation) reflect important ecological contrasts (see below) but do *not* form an evolutionary sequence: not only are all extant, but it is nonsense to talk of a species group (e.g. Old World monkeys) as ancestral to a single species (humans). The progression here is, in fact, an archaic (pre-evolutionary) hierarchic view of natural diversity given an evolutionary gloss.

Dissatisfaction with grades and the continuous search for ancestors has led some palaeoanthropologists to advocate a cladistic approach to hominid phylogeny. Cladistics is based upon the principles distinguishing positive and derived characters summarised above, but these are more formally explicated and rigorously applied, as set out by Hennig (1966). The major contribution of cladistics has been to emphasise the need to distinguish levels of homologous similarity: mere totting up is inadequate because of differential evolution rates. Two groups may share many similarities yet only distant ancestry if both are slowly evolving, while a recently differentiated form may yet be markedly different if it is evolving rapidly: overall similarity here obscures the real sequence of splits. Cladists therefore argue that phyletic reconstruction is best approached via the branching sequence (i.e. the succession of speciation events). The sequence is deduced from a comparison of character states across the forms considered, and the determination of polarity changes in such characters (i.e. primitive to derived).

In cladistics, groups are successively defined by sets of shared derived (synapomorphous) characters until a single derived (autapomorphous) character, uniquely defining a lineage, is identified; primitive (plesiomorphous) traits do not denote any especially close relationship. In the primate example above, bilateral symmetry, homoeothermy and a vertebral column are plesiomorphous for all mammals, and so reveal nothing of primate interrelationships. The possession of an opposable hallux is synapomorphous for primates as opposed to non-primates among

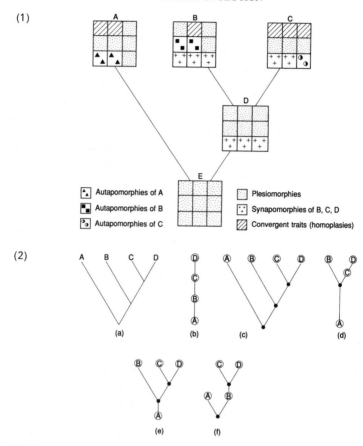

Figure 2.2 (1) Cladistic analysis and components of affinity. Ancestor E gives rise to species A and D. The latter subsequently speciates to produce B and C. Summing resemblance suggests that A and C are more closely related (5/9 traits similar) than B and C (4/9). However, A and C similarities are due to convergence (homoplasy) or primitive retentions from E. Identification of derived character sets common to B, C and D but lacking in A (crosses) reveals the actual pattern of relationships and shows B and C to be more closely related to each other than either is to A (see also text). Modified from Martin (1990). (2) (a) Cladogram with 4 groups and (b)–(f) different phylogenies compatible with the cladogram. Blacked circles: hypothetical common ancestor.

mammals, and bilophodont molars autapomorphous for cercopithecoids (Old World monkeys) as opposed to other primates. For any investigation of phylogeny within Old World monkeys, bilophodanty then becomes plesiomorphous, so that character states are relative, not absolute.

The result of cladistic analysis is a nested set of groups defined by progressively derived characters. These are expressed as a cladogram, a tree with a sequence of branching points but lacking an absolute or relative time-scale, and with no ancestor–descendant sequences, although close-related (sister) groups share a recent (but unknown) common ancestor. There is nothing inherently evolutionary about a cladogram; it could just as appropriately be utilised by creationists, and it is perhaps best viewed as summarising a particular interpretation of patterns of organic diversity. The cladogram may, however, form the basis for phyletic schemes that are consistent with its branching sequence. There are usually several phy-logenies compatible with the cladogram, and additional information (dating, zoogeography, etc.) may be used to decide which of these is the most plausible. However, some workers refuse to extend their analysis beyond the construction of the cladogram since in their view to do so entails a loss of rigour by introducing a subjective element into the analysis.

For cladists, degrees of monophyly are identified by successive sets of synapomorphies until individual lineages defined by their autapomorphies are reached. The groups recognised by cladists cannot be ancestral to any others: the presence of at least one autapomorphy (which, by definition, is unique to that group) prevents this. What can be established are sister groups (those that are each other's closest relatives because they possess most synapomorphies) sharing an (unknown) immediate common ances-tor. Cladists stress that their method is logical, internally consistent and testable and, as such, superior to the evolutionary systematists' approach, which they consider to lack a rigorous base and to be too preoccupied with ancestor–descendant relationships that are impossible to demonstrate from the fossil record.

Why then are ancestral groups so widely recognised in the fossil record? Cladists consider that so-called ancestral groups are not monophyletic. For them, monophyletic groups not only share a common ancestor but contain *all* descendants of that ancestor. What frequently happens in evolutionary systematics, they argue, is that some markedly divergent descendant forms are removed from the rest of the group and separately identified. What is then left behind is a paraphyletic group that does not contain all the descendants of a common ancestor, and which is characterised by shared primitive features (plesiomorphies) but lacks defining derived characters. Paraphyletic groups include Invertebratae (lack of a backbone is not a defining character), Pisces (fish; terrestrial vertebrates removed), Reptilia (absence of mammal and bird defining features) and, nearer home, apes (see below): in other words most of the favoured ancestral groups of traditional systematics. Their great disadvantage is that since they lack defining characters they are artefacts rather than real (Patterson, 1982).

Monophyly has traditionally referred to a taxon, all members of which are descended from the nearest common ancestor; the paraphyletic groups of cladists can be monophyletic for evolutionary systematists since the latter do not require *all* descendants of the ancestor to be included for the group to be monophyletic. In fact, cladists have been severely criticised for applying the term monophyletic to descent, so that a taxon is recognised not by its characters, but by what it evolves into. Ashlock (1971; 1979) advocated retaining the traditional use of monophyly and proposed the term holophyly for a group including all descendants of the nearest common ancestor (the cladists' sense of monophyly).

The other category of non-natural groups is those that are polyphyletic, i.e. grouped by homoplastic (analogous) features. A group of winged animals (birds, insects, pterodactyls, bats, etc.) is an example. In practice, these are more easily recognised and avoided and therefore less troublesome in most investigations than paraphyletic groups.

Cladistic analysis reconstructs genealogy or origination of lineages, but cannot accommodate change within lineages, so that for cladists chronospecies do not exist. Each clade consists of just one species, with a real point of origination (at speciation) and of termination (at extinction or subsequent speciation when new daughter species arise). There is an obvious parallel with the punctuational view of evolution here. And yet both the fossil record and field studies of modern species indicate that anagenetic change occurs.

Nor can cladistic analysis accommodate degrees of lineage divergence resulting from different evolutionary velocities. In contrast, for evolutionary systematics, degree of divergence may be more significant than recency of common ancestry (Mayr, 1981). Birds and crocodiles are a case in point: these are sister groups, with crocodiles more closely related genealogically to birds than they are to other reptiles. Since differentiation, crocodiles have changed little (developed few autapomorphies) compared with other reptiles, whereas birds have evolved many derived features: the adaptive shift to an aerial zone has resulted in markedly greater changes in one sister group than in the other, and so they are generally referred to separate grades. Grades are particularly useful for reflecting adaptive reorganisation in this way but, as Mayr (1981) notes, 'the cladist virtually ignores this ecological component of evolution'.

Another way of looking at this, as Mayr points out, is to consider the treatment of shared and uniquely derived characters. Cladists are interested in the pattern of synapomorphies with which to reconstruct the branching sequence, and autapomorphies are the limiting cases which identify the end products of this process, the individual lineages. Evolutionary systematists are also interested in the number and significance of autapomorphies relative to synapomorphies since they may reflect important adaptive shifts. This is why their approach differentiates between crocodiles and birds, and between African apes and humans (see below).

The impact of cladistics has produced undoubted benefits in promoting awareness of the need to set out the assumptions and procedures adopted in phyletic reconstruction, and to distinguish between different levels of homologous similarity. However, the inability of cladistic analyses to incorporate change within lineages and degrees of phyletic divergence, as opposed to the sequence of divergent episodes, greatly reduces the information content of the analysis, as does the rejection of any ancestor–descendant relationships.

The claims made for cladistics have not gone unchallenged. Critics (e.g. Simpson, 1975; Gingerich and Schoeninger, 1977; Gingerich, 1978, etc.; Hull, 1979; Trinkaus, 1990) argue that its assumptions are no less sweeping, involve circular reasoning and provide only an incomplete and biased view of phylogeny. Cladistics depends critically upon the identification of character states and their directions; these are most reliably determined in groups where there are large numbers of closely related forms, so that extensive comparisons can be made; it is much more difficult when the group is represented by only a single living species (as with *Homo*). Uncertainties over the most appropriate 'out-group' comparators may distort the polarity clines derived, and fossil forms must be incorporated into the analysis at an early stage, despite the dangers that they pose for polarity determination (see above).

It should also be noted that, while cladistic methods may show previously defined groups to be heterogeneous (para- or polyphyletic), cladistic principles do not aid in constructing the groups that enter cladistic analysis. Indeed, the boundaries and content of the individual groups will largely predetermine the output of the analysis. This is of particular concern with hominids, where the status of individual fossils may be uncertain or disputed, yet their assignment will influence the study's outcome. For maximum reliability, therefore, a phenetic study to construct groups should logically precede cladistic analysis, which can then be used to test the monophyly of the phena and derive additional information about the distribution of character states across the groups (see also Corrucini and McHenry, 1980; Bilsborough and Wood, 1986; 1988).

Controversies about how evolution proceeds and about methods of phyletic inference, i.e. how best to reconstruct past evolutionary events whatever form they might have taken, are logically distinct. However, it should be fairly obvious that the two separate debates have implications for each other: those convinced by the punctuationist interpretation are likely to be predisposed towards cladistic methodology with its emphasis on a rigorously determined sequence of splits, whereas phyletic gradualists are more likely to favour classical stratophenetic approaches that also document anagenesis and ancestor–descendant sequences.

2.5 Classification

The main contrasts between cladists and evolutionary systematists are highlighted by their approaches to biological classification. Cladists base their categorisation entirely upon genealogy, arguing that only monophyletic (i.e. holophyletic) groups should be classified, and that sister groups should be of equal rank. The cladogram is converted directly into the Linnean classificatory system, with the earliest splits in the sequence having the highest taxonomic rank and the most recent the lowest. Evolutionary systematists argue that the classification should be compatible with inferred phylogeny but need not be closely determined by it and may also reflect other factors such as adaptive contrasts and degrees of divergence; groups with a recent common ancestor may be distinguished at a relatively high taxonomic level if they are markedly different. These contrasting approaches can be illustrated through their arrangement of the highly relevant examples of apes and humans.

A well-known and influential classification of mammals on evolutionary systematic principles is that of Simpson (1945). This is turn serves, whether in its original form or modified by other workers, as the basis for the best-known arrangement of primates. In Simpson's scheme the order Primates is divided into two suborders: Prosimii (lemurs, lorises, bush babies and tarsiers) and Anthropoidea (monkeys, apes and humans). Within the Old World Anthropoidea, or catarrhines, the major groups are the superfamilies Cercopithecoidea (Old World monkeys) and Hominoidea (apes and humans). The apes are divided into lesser apes (gibbons and siamangs) and great apes (chimpanzee, gorilla and orang). Simpson included all these within the family Pongidae, recognising two subfamilies: Hylobatinae (lesser apes) and Ponginae (great apes). It is now more usual to distinguish them as separate families, the lesser apes as Hylobatidae and the

Pongidae reserved for the great apes. In both the original scheme and its modifications, humans, their immediate ancestors and relatives are placed in a separate family, Hominidae.

In recent years a good deal of (mainly) molecular evidence has accrued to show that chimp, gorilla and humans resemble one another more closely than do the African apes and orang, which are markedly distinct in many characters (see chapter 4). This leads to the conclusion that chimp, gorilla and humans shared a common ancestor more recently than did the African apes and orang, and that the marked morphological contrasts between humans and apes result from rapid evolution within the hominid clade after that final split, rather than that the hominid–pongid split predates division of the African and Asian apes. In other words, just as the African apes are our closest living relatives so we are theirs, and grouping chimp, gorilla and orang in the family Pongidae with humans in Hominidae does not accurately reflect the genealogical relationship.

Evolutionary systematists recognise this but have no difficulty living with the situation since the three apes represent the same adaptive grade. The major contrasts between African apes and humans, reflected in such hominid autapomorphies as truncal erectness, bipedal locomotion and masticatory systems (see chapter 3), are, for them, of greater significance than the (mainly molecular) African ape-human synapomorphies.

Cladists, on the other hand, find this arrangement unacceptable because Simpson's family Pongidae is paraphyletic, hominids having been hived off into a separate family. They advocate retaining Pongidae for the orang only, and group chimp and gorilla with humans within Hominidae to reflect more accurately the inferred genealogical relationships of the species (Bonde, 1977; Andrews and Cronin, 1982; Patterson, 1982). However, this requires a redefinition of Hominidae and 'hominid' since it destroys the traditional use of the taxon as reserved for humans and their immediate, exclusive ancestors and requires statements along the lines of 'Bipedalism is characteristic of all hominids except chimp and gorilla' (Martin, 1990).

Neither classificatory approach is 'correct' or 'incorrect'; both are legitimate and designed to express different aspects of evolutionary diversity. Classifications serve several purposes: they act as summaries or memory aids (i.e. data storage) and also organise information in ways that promote further generalising from new studies. To achieve these ends the classification must allow easy information retrieval and be an effective communicatory device—the appropriate test to apply to any particular scheme is a practical, utilitarian one of whether it achieves these goals.

Incorporating chimp and gorilla within Hominidae requires redrawing

Table 2.1 Some taxonomies of living apes and humans. The differences between pairs (a), (b) and (c), (d) illustrate the contrasting approaches of grade- and clade-based classifications, with the former reflecting degree of divergence, the latter recency of common ancestry. The differences within each pair, (a) versus (b) and (c) versus (d), illustrate that taxonomy is a subjective art, not an exact science.

Grade-based		Clade-based	
(a) (Simpson, 1945)	(b) (Conroy, 1990)	(c) (Andrews and Cronin, 1982)	(d) (Andrews, cited in Aeillo and Dean 1990)

Superfamily: Hominoidea
Family: Pongidae
	Superfamily: Hominoidea
	Family: Hylobatidae
		Superfamily: Hominoidea
		Family: Hylobatidae
		  Hylobates
			Superfamily: Hominoidea
			Family: Hylobatidae
			  Hylobates

Subfamily: Hylobatinae
  Hylobates
  Symphalangus
	Subfamily: *Hylobates*
	Family: Pongidae
		Family: Pongidae
		  Pongo
			Family: Hominidae

Subfamily: Ponginae
  Pongo
  Pan
  Gorilla
	Subfamily: *Pongo*
	  Pan
	  Gorilla
		Family: Hominidae
		Subfamily: Gorillinae
			Subfamily: Ponginae
			  Pongo

Family: Hominidae
  Homo
	Family: Hominidae
	  Homo
		Tribe: *Pan*
		  Gorilla
		Subfamily: Homininae
		  Homo
			Subfamily: Homininae

			Tribe: Gorillini
			  Pan
			  Gorilla
			Tribe: Hominini
			  Homo

all our concepts of a 'hominid' and a 'hominid trait' and obliterates the structural and adaptive contrasts between the African apes and ourselves. Many, including the author, consider this too great a price to pay for genealogical purity, and in the rest of this book the family Hominidae and the terms 'hominid', 'hominid character', etc. are used in the traditional sense of evolutionary systematics as they are, for example, in Simpson (1945), Conroy (1990) and Martin (1990).

2.6   The temporal framework

One of palaeoanthropology's long-term goals has been to establish the time-scale for human evolution. For a long time the only chronology

Table 2.2   Some relative dating methods. Most are only reliable within individual sites. Even when possible, between-site correlations can usually only be made with confidence on a very local basis.

(1) Stratigraphy	Based on the principle of geological superposition i.e. deposit at bottom, youngest at top, except where faulting, etc. obviously disturbs the ordering. If distinctive strata are identified ('marker beds') it may be possible to cross-relate two or more sites into a general sequence. Geochemical 'finger-printing' of individual deposits, e.g. in East Africa (chapters 5 and 6), has refined this approach.
(2) Fauna	Where the patterns of animal evolution are well known, distinctive faunas may allow sites to be placed within a time span. Rapidly evolving groups are generally more useful than slowly evolving ones, e.g. pigs, monkeys and elephants in the Plio–Pleistocene of East Africa, and microtine rodents in Pleistocene Eurasia.
(3) Flora	Vegetation changes, usually revealed by fossil pollen, may also aid ordering, e.g. late Pleistocene and Holocene sites in northern Europe.
(4) Climate	Where distinctive climatic episodes are recognised, climatic evidence may aid in relative placement. Often used with (2) and (3) which provide much of the relevant evidence; other data are geological. Traditionally used to construct a sequence for European sites based upon a glacial/interglacial cycle. However, this has been undermined by the realisation that there were very many more oscillations than previously thought (chapters 7 and 8).
(5) Chemical	The chemical composition of organic material (e.g. bone) may provide a relative age for a specimen. Indicator elements include uranium and fluorine (both increase with age) and nitrogen content (decreases with age). However, levels usually apply only to the sequence at a single site, and it may not be possible to cross-relate sites.
(6) Archaeological	Assumes that particular artefacts (especially Palaeolithic tool kits) follow a temporal sequence, which is only correct in the most general sense: in most cases tool forms are determined by functional, cultural and technological considerations, not age. Of very limited application.

possible was a relative one that ordered specimens into an older–younger sequence using the traditional techniques of stratigraphic geology (Table 2.2). However, during the last 40 years major advances have been made in developing an absolute chronology, measured in years. At first only later human evolution was so calibrated, but appropriate techniques now provide dates for the origin and earliest phases of hominid evolution, so that while gaps, uncertainties and omissions remain the broad temporal framework is clear, increasingly allowing for recognition of periods of rapid change and others of stasis.

Many absolute dates are based upon radiometric methods which involve the decay of an unstable radioactive isotope of an element to its stable inactive form. The element's decay is usually exponential, so that while the absolute amount varies over time the rate, expressed by the half-life, is constant and known. Various dating methods are in use of which the more important are summarised below. Fuller accounts include Aitken (1985; 1990) and Parkes (1986).

2.6.1   Carbon-14 (radiocarbon)

The first radiometric method to be developed, and the best known, measures carbon-14 (^{14}C), which occurs as a minute proportion of all carbon in the atmosphere in the form of carbon dioxide, and is incorporated into plants and animals through the food chain. After death ^{14}C uptake ceases, and the decay of ^{14}C to nitrogen relative to the amount of 'ordinary' carbon, ^{12}C, gives the specimen's age.

Decay is relatively rapid (the half-life is < 6000 years) so that conventional ^{14}C dating is limited to the last 40 000 years or so; the new technique of mass accelerator dating (Hedges and Gowlett, 1986) promises to extend this to 80 000–100 000 years, but even so it will only be applicable to later human evolution. In the conventional method the ^{14}C concentration is estimated by measuring the decay of ^{14}C atoms over 1 or more days; at ages of $\geqslant 40000$ years the ^{14}C radioactivity is indistinguishable from ordinary background radiation. Moreover, the decay is random and so extrapolation may under- or overestimate the total ^{14}C present ('counting error'). The mass accelerator method measures a proportion of total ^{14}C (not just the fraction presently decaying) by energising the sample and separating ^{14}C ions from others. This gives greater precision, allows the use of much smaller samples and potentially extends the dating range once contamination by minute traces of modern carbon is overcome.

Other variables affect the accuracy of ^{14}C dates. For consistency the early half-life estimate of 5568 years continues to be used instead of the more accurate 5730 years, so that ^{14}C years are shorter than calendar years: a factor of c. 1.03 is needed to transform the former into the latter. The rate of ^{14}C production, once assumed to be constant, is now known not to be so, and a correction factor must be applied depending upon the specimen's uncorrected age. Experimental and other errors must also be taken into account. Within these constraints, ^{14}C is a sensitive and widely applicable dating technique and has produced a detailed time-scale for the final Pleistocene.

2.6.2 Potassium–argon (K–Ar) and ^{40}Ar–^{39}Ar

This depends upon the decay of potassium-40 (a small but constant proportion of all potassium) to calcium-40 and then to the inert gas argon-40. The ratio of ^{40}K to ^{40}Ar indicates the age of the specimen. The rate of decay is very slow and the half-life correspondingly long (1250 million years), so the method is applicable to the earlier phases of hominid evolution but not the later, because since then so little potassium has decayed that it cannot be measured accurately. Recent technical developments promise applications down to c. 30 000 years (Aitken, 1990) but these are not yet widespread. The technique also requires that all argon produced by the decay is retained, and that no other (atmospheric) argon is introduced into the sample. These considerations limit its use to igneous (volcanic) rocks, either tuff (ash) or lava, but the fossils themselves cannot be dated so that the specimen's context in relation to the datable igneous rock needs to be secure. The technique has been most widely used in East Africa, where many fossil specimens are associated with, or bracketed by, volcanic episodes of Rift Valley formation (see chapters 4–6). The amounts of potassium and argon are measured by different techniques on separate parts of the sample, on the assumption that they are evenly distributed throughout, which may not be correct. In order to overcome this difficulty, ^{40}Ar–^{39}Ar dating was developed.

In the ^{40}Ar–^{39}Ar method the sample is irradiated in a nuclear reactor to convert ^{39}K into ^{39}Ar so the latter can be used to measure the former, from which the proportions of ^{40}K can in turn be calculated. This method therefore measures the two Ar isotopes, using the same techniques on the same portion of sample, rather than the potassium–argon ratio. It thus in theory overcomes one difficulty of 'conventional' potassium–argon dating and has been used, for example, at Koobi Fora (East Rudolf) (see chapter 6). However, results have not always been consistent and some ^{40}Ar–^{39}Ar

applications have been subject to disturbing variation because minute errors in measuring the argon quantities can apparently affect the ages derived. In addition, various reheating events, associated with multiple volcanic episodes, may give misleading results.

Most potassium–argon dating applications are for *c.* 0.4 million years or more. Carbon-14 is only applicable to specimens *c.* 0.05 million years old, with the promise of up to *c.* 0.1 million years and there is thus a significant interval between these limits for which reliable absolute dates are not available. Various techniques promise to bridge the gap but are currently of limited applicability. Some of the more important of these techniques are described in the following sections.

2.6.3 *Thermoluminescent (TL) and electron spin resonance (ESR) dating*

Crystalline materials such as rocks, calcite and pottery are subject to natural radiation, mainly from uranium-238 and -235, thorium and potassium-40 within the material itself or the surrounding deposit. The radiation causes release of electrons, and a proportion of these are trapped by lattice defects in the material's structure. Heating releases the electrons, which then combine with anions to emit a pulse of light—thermo-luminescence. Heating in prehistory (e.g. in a hearth) will release trapped electrons to date and set the electron 'clock' back to zero; those released and trapped thereafter will depend upon the average radiation intensity and the time elapsed since the clock was reset. If the radiation rate is known, then laboratory reheating and measuring the light produced by the freed electrons can be used to date the specimen's initial heating.

This technique was originally used for dating recent pottery, but it has also been applied to Upper Pleistocene burnt flint, sandstone and limestone, and loess deposits (Aitken, 1985; 1990).

Electron spin resonance (ESR) also allows calculation of the trapped electrons, radiation dose and thus age by measuring the wavelength range and amount of energy absorbed when the sample is placed in a magnetic field. Since the method obviates the need for heating, it is potentially especially valuable for dating teeth and calcite but application to bone has been less successful. Uranium uptake by organic materials continues after death and this will influence the rate of free electron production. Two age estimates are usually given, therefore, depending on whether uptake is assumed to have occurred shortly after death (early uptake (EU) date) or at a constant rate since death (linear uptake (LU) date). Full accounts are given in Aitken (1990) and Grun and Stringer (1991). Recent applications of both TL and ESR techniques to late Middle/Upper Pleistocene sites have significantly modified previous notions of biological and technological change (see chapters 7 and 8).

2.6.4 *Amino acid racemisation dating*

Amino acid racemisation dating of bone depends upon the gradual change of one structural form of aspartic acid to its mirror image form, or enantiomer. Organisms produce only L-aspartic acid, but after death this progressively changes to the mirror image D-form, by the process of racemisation. The rate of change is temperature dependent, so that dating techniques can only be applied to bone from constant environments such as deep cave interiors. The racemisation rate can be determined from the temperature of the deposit, and in principle the age can then be calculated from the ratio of D to L enantiomers. In practice, estimated ages can vary erratically compared with, for example, carbon-14 dates, and errors are often high. The technique has therefore not been much used in recent years, although in the 1970s it was applied to a number of Middle and Upper Pleistocene specimens.

2.6.5 *Uranium series dating*

Calcium carbonate (calcite) in the form of travertine, stalagmite, etc. generally incorporates uranium, but not the more insoluble thorium, during its formation. However, radioactive uranium-234 decays to thorium-230 with a half-life of 75 400 years, so if the initial concentration of uranium is known the uranium–thorium ratio gives the age of the calcite. Other isotopes of uranium and its daughter products (such as protactinium) can in principle be used for dating, but in practice the $^{234}U-^{230}Th$ ratio is commonly used. This method has been applied to Middle and Upper Pleistocene (50 000–500 000 year old) specimens, and is principally used on corals and calcite from cave/rock shelter contexts, e.g. encrusting or sealing in a specimen. Direct dating of bone has also been attempted, although with less success.

2.6.6 *Palaeomagnetism*

Although not strictly speaking a dating method, palaeomagnetic data have proved valuable in conjunction with other methods by providing cross-checks, or localising a specimen within a particular time zone. The last 3.5

million years (my) have seen several abrupt changes in the direction of the earth's magnetic field, so that the north magnetic pole moves to the south and vice versa. Several such magnetic chrons or epochs (lasting $\geqslant 0.5$ my) are recognised, within which are numerous shorter subchrons or events (lasting $\leqslant 0.1$ my), when the field also switched. The direction of the magnetic field at the time of a rock's formation may be preserved, thus

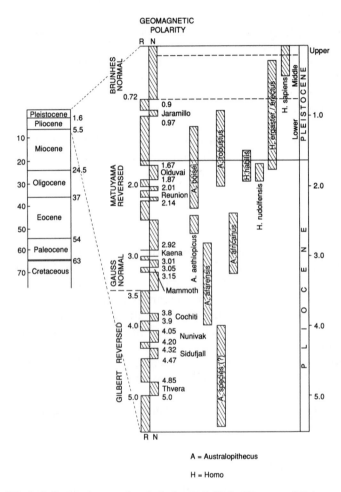

Figure 2.3 Left: Caenozoic era and geological periods. Right: Pliocene and Pleistocene time-scale with geomagnetic record (chrons and subchrons) at left and main hominid species to right. Ages in million years ago (mya).

Table 2.3   The palaeomagnetic column: Plio–Pleistocene polarity switches, chrons and subchrons.

Chron (epoch)	Subchron (event)
Brunhes (N) (present–0.72 my)	
Matuyama (R) (0.72–2.48 my)	Jaramillo (N) (0.90–0.97 my)
	Olduvai (N) (1.67–1.87 my)
	Réunion (N) (2.01–2.04 my)
	(2.12–2.14 my)
Gauss (N) (2.48–3.40 my)	Kaena (R) (2.92–3.01 my)
	Mammoth (R) (3.05–3.15 my)
Gilbert (R) (3.40–5.28 my)	Cochiti (N) (3.80–3.90 my)
	Nunivak (N) (4.05–4.20 my)
	Sidufjall (N) (4.32–4.47 my)
	Thvera (N) (4.85–5.00 my)

N = normal; R = reversed; my = million years.

indicating whether it is normal or reversed, and other evidence may help determine the exact magnetic episode within which it falls. Table 2.3 summarises the main polarity episodes relevant to hominid studies.

All of the above techniques give valuable information on the chronology of human evolution, but all are subject to error factors, and the most reliable dates are those where multiple lines of evidence converge on a single figure. Such cross-disciplinary studies are best known from East Africa, especially at sites such as Olduvai Gorge (Tanzania), Koobi Fora (Kenya) and the Hadar and Omo sites (Ethiopia) (chapters 5 and 6). Nor do such 'high-tech' absolute methods mean that traditional relative dating approaches must be relegated to a minimal role: at Koobi Fora faunal analysis and detailed stratigraphic correlation, aided by geochemical 'fingerprinting' of volcanic tuffs, led to major revision of initial ^{40}Ar–^{39}Ar dates. Cross-checks that yield dates compatible with all lines of evidence remain the most reliable. Figure 2.3 summarises major divisions of the time grid for human evolution obtained using the above and other methods.

CHAPTER THREE

HOMINID STRUCTURE AND FUNCTION

3.1 Introduction

This chapter summarises functional aspects of some of the distinguishing human features noted earlier: erect posture and bipedal locomotion; manual dexterity and manipulation; masticatory activity and food processing; and brain expansion and behavioural complexity.

Space allows only a brief outline of the more significant features evident in the hominid fossil record; more detailed information on structural–functional aspects can be found in any standard textbook of human anatomy. Campbell (1966) provided a pioneering survey of functional systems in human evolution, while Napier (1980) and Passingham (1982) are examples of accounts focusing on particular systems. Broader perspectives from primate evolutionary biology are provided by Fleagle (1988), Conroy (1990) and Martin (1990), while the most comprehensive and up-to-date account of hominoid functional anatomy in a phyletic context is that of Aeillo and Dean (1990).

3.2 Biomechanics

Many of the distinctive skeletal features of hominids can be interpreted biomechanically as adaptations to extend the range of movement about a joint (e.g. at the base of the thumb) or increase the efficiency of muscles involved in particular actions (e.g. moving the thigh on the hip). Complex movements can generally be reduced to their simpler components of flexion, extension, abduction, adduction and rotation. Broadly speaking, flexion involves bending a structure at a joint (e.g. the arm at the elbow, the leg at the knee), whereas extension straightens it. Abduction moves the structure away from, and adduction towards, the midline of the body. Rotation, as the name implies, rotates a structure: the skull on the neck, twisting movements of the vertebral column, and the forearm at the elbow are all examples.

The limbs and other body structures can be analysed as levers, with the weight they support as the load, the joint as the fulcrum, and the power provided by muscle contraction. The muscles' efficiency is improved by moving them away from the joint about which they act, so increasing the power or moment arm on the 'see-saw principle': a smaller force applied some distance from the pivot is more effective than a larger force applied close to the pivot. The olecranon process at the elbow, the proportions of the hominid hip bone and the prominent human heel are all examples of structures that improve muscle efficiency by moving them away from the fulcrum, so lengthening the lever arm (see below).

3.3   Head and skull

Cranial remains figure prominently in the hominid evolutionary record for several reasons. Taphonomic factors undoubtedly play a part: heads contain little meat and so are unattractive to carnivores and scavengers. In addition crania are easily recognised compared with broken long bones, whilst in evolutionary terms hominid cranial changes have been spectacular, involving face and jaw reduction and brain expansion. Cranial features therefore serve both to characterise and identify various hominid species, and also provide much information on their ecology and adaptation.

The head serves multiple functions: it houses most of the sense organs and the main structures of the central nervous system. It incorporates the food processing apparatus and also provides means of communication, both verbal and non-verbal, between individuals. Corresponding to these several functions, head and skull may be conveniently if roughly divided into the face, housing sense organs and jaws, and the neurocranium or braincase. The basicranium, where the skull articulates with the vertebral column and where the mandible attaches, is relatively invariant, and brain expansion has primarily affected the cranial vault rather than the base.

The main bones of the face are the mandible (lower jaw) and the maxilla (mid-upper face and most of the palate). Anterior to the maxilla in other primate species, including early hominids, is the premaxilla, bearing the incisor teeth. The ossification centre of the premaxilla is also evident in the human fetus, but before birth is overgrown by the maxilla and no longer distinguishable as a separate bone. On either side of the maxilla are the

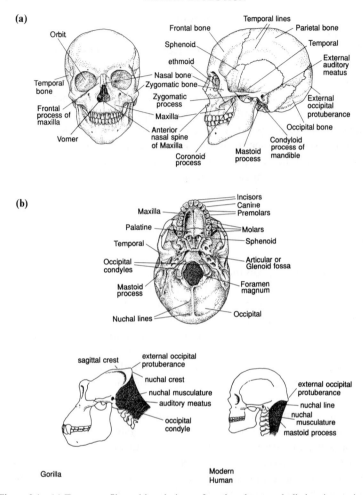

Figure 3.1 (a) Front, profile and basal views of modern human skull showing major bones and main features. (b) Gorilla and human skulls showing contrasts in face and jaw proportions, articulation with vertebral column, and size and extent of nuchal muscles.

zygomatic (cheek) bones, running back and linking face to braincase. Face and braincase also meet where the maxilla and zygoma suture with the frontal. The bones of the vault—frontal, two parietals forming the sides of the vault with the temporals immediately below, and the occipital at the rear—are large, curved and smooth and thin in section. The temporals and occipital also extend onto the cranial base, meeting just in front of the foramen magnum where the spinal cord and blood vessels pass into the brain. On either side of this are the occipital condyles, where the skull balances on the vertebral column, and the glenoid fossae, where the mandibular condyles attach.

Variation in the sense organs is generally relatively minor among catarrhines and most contrasts in facial size and proportions reflect contrasts in the jaws and dentition.

3.4 Dentition

Broadly speaking the anterior teeth, incisors and canines, are slicing teeth, serving to chop food items prior to chewing and swallowing, while cheek teeth, premolars and molars, crush and grind food, so increasing the surface exposed to enzyme action. Each tooth is mainly composed of a moderately hard substance, dentine, over which there is a much harder coating of enamel. In hominids and some Miocene apes the enamel coating is thicker than usual for primates, suggesting a tough, resistant diet, whereas enamel appears secondarily thinned in the African apes. This may be an adaptation for eating shoots and other tree products since the dentine, once exposed through the thin enamel, wears more rapidly, so producing a chisel-like edge at their interface.

All catarrhines share a common dental formula (I 2/2, C 1/1, Pm 2/2, M 3/3), but there are major contrasts in the form of the tooth row, and in the size and structure of individual teeth. In other primates the dental arcade resembles an inverted U, whereas in hominids the tooth row is more evenly curved, approximating a parabola. The incisors are flat and blade-like with a slicing action; they are particularly large in frugivorous forms such as the chimp and orang, and while only moderately sized in most hominids are larger in some very early specimens (chapter 5). In most primate species canines are conical, piercing teeth, projecting well beyond the tooth row and with upper and lower canines shearing against one another, adding a slicing action to the piercing component. Their projection results in a diastema (gap) in the upper jaw between lateral incisor and canine for the tip of the lower canine, and in the mandible between lower canine and anterior premolar for the upper canine's tip. In contrast, hominid canines are small and chisel-like rather than conical and, since they are non-projecting, form an extension of the incisive cutting edge.

The bicuspid premolars are partly cutting, partly grinding teeth. In forms with large canines the first lower premolar is sectorial, i.e. one cusp is very large and the curved anterior face shears against the back of the upper canine as the jaw closes. In hominids both premolars are similar in form and non-sectorial, but there is a tendency for additional cusps (molarisation). The larger molars are multicusped grinding teeth with a complex topography so that the upper molar cusps fit into hollows on the lower molar surface, producing a grinding action when the jaw closes, rather like a morter and pestle, an action enhanced in hominids by their distinctive rotary chewing pattern. This combination of vertical and lateral movement wears the molar surface smooth relatively rapidly; thicker enamel clearly extends the functional life of teeth subject to such heavy attritional wear. Even so, some early hominids show spectacular tooth crown destruction (chapter 5).

Teeth have also been used to study individual (ontogenetic) age and development in early hominids. Early studies concentrated on the eruption pattern of the dentition, sometimes combined with estimates of tooth wear in older individuals. However, there are major difficulties with this approach, quite apart from the uncertainties related to estimating rates of tooth wear. Modern ape and human studies indicate that there is considerable variability in tooth eruption timing and patterns: both duration and sequencing of individual teeth can vary a good deal, so that caution is required in assessing specimens. Nonetheless, some general contrasts are apparent: pongid incisors and canines are large, grow for a longer period and erupt relatively late (around the same time as, or later than, the second molars), whereas in humans the anterior teeth are small and erupt relatively early (around the time of the first molar) while second molar eruption is delayed and that of the third molars particularly so.

Immature early hominid fossils can be assessed in this way, but the necessity to assume a particular model based on either modern ape or human standards for the timing and sequencing of eruption is a major drawback. As distinctive forms it is inherently unlikely that early hominids closely matched either modern apes or modern humans in their developmental patterns.

Fortunately studies of tooth microstructure can help calibrate dental development in fossil specimens. Electron microscopy reveals tooth enamel to be laid down incrementally, with multiple cross-striations and coarser striae of Retzius that outcrop at the tooth surface as ridges (perikymata). The underlying mechanism for these is unknown, but cross-striations appear to be laid down daily and striae of Retzius every 7–9 days (a circaseptan). Counts of cross-striations and striae of Retzius can therefore be used to time the period of crown formation of individual teeth and, if

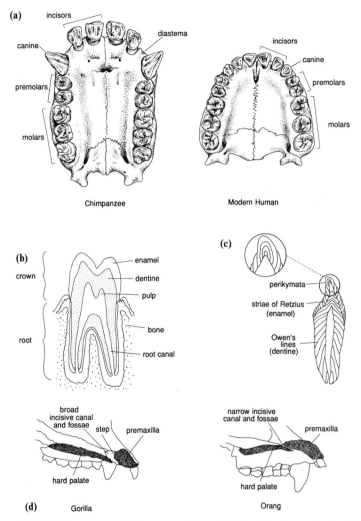

Figure 3.2 (a) Chimpanzee and human maxillary dentition showing contrasts in tooth morphology and proportions and shape of dental arcade. (b) Section of molar tooth showing dental tissues. (c) Incremental growth in dental tissue. Regular bands (striae of Retzius) are apparent in enamel, forming ridges (perikymata) where they outcrop at the surface. Owen's lines reflect incremental growth in dentine. (d) Cross-sections of palate and nasal aperture in gorilla and orang showing contrasts in structure between African and Asian apes (see text). (c) taken from Aeillo and Dean, 1990; (d) taken from Ward and Kimbel, 1983).

several teeth are preserved in the specimen, the period over which dental formation occurred and the timing of particular events such as canine or first molar eruption. In this way information on both pattern and duration of early hominid tooth development can be obtained, and age of death in infants and juveniles estimated (Bromage and Dean, 1985; Dean, 1987; Beynon and Dean, 1988; Aeillo and Dean, 1990). The results have already led to revised views on early hominid maturation. They indicate relatively short growth periods similar to apes rather than humans, but with unique development patterns, unlike modern hominoids. They also reveal major contrasts between early hominid species, especially in the growth of the anterior teeth (see chapter 5).

Tooth microstructure may also record trauma such as birth or disease, and dental development in primates correlates strongly with brain growth and several life history variables (Smith, 1989) so it should be possible to generalise from dental evidence to other aspects of individual development and life history such as age at sexual maturity, reproductive patterns and lifespan. The findings noted above have not gone unchallenged, and many uncertainties remain (Mann *et al.*, 1990), but tooth microstructure offers the potential for much valuable information on early hominid ecology and development.

3.5   Masticatory activity

The lower jaw acts as a lever, raised by muscles in food processing. The teeth are set in the horizontal mandibular body (corpus), while the masticatory muscles attach to the vertical ramus, which also bears the condyle where the mandible articulates with the cranium. In hominids the face is deep and the teeth's occlusal surface lies well below the pivot, which has several consequences: muscle force is directed vertically over the cheek teeth so that maximum power is generated here, and these teeth have an antero-posterior component added to their action as the jaw closes, so increasing grinding efficiency. In large-jawed forms with a tough diet the mandible has to resist marked compressive force, and the corpus tends to be deep, especially below the molars. In hominids the jaw also has to withstand lateral shearing forces generated in rotary chewing and the corpus is shallower but thicker, with lateral buttressing running along the inside. The chewing forces converge on the symphysis—the middle region of the front of the jaw—and this is accordingly reinforced. In early, large-jawed hominids, reinforcement consists of two internal ridges of bone (the superior and inferior transverse tori), but in modern small-jawed humans there is an external buttress—the chin.

Gravity ensures that the jaw hangs open: closing it involves several muscles. Temporalis is a large, fan-shaped muscle originating on the sides of the braincase and inserting on the mandible's coronoid process and anterior ramus. In frugivorous primates the fibres run obliquely, so exerting

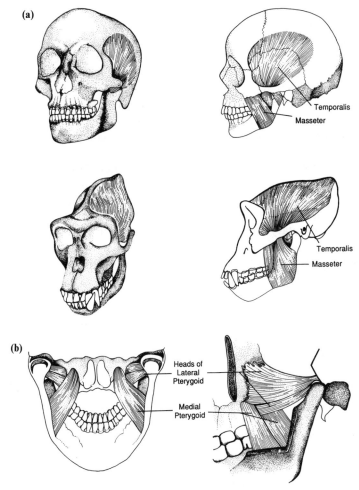

Figure 3.3   (a) Three-quarter and profile views of human and gorilla skulls to show contrasts to size and orientation of masticatory apparatus. (b) Pterygoid muscles in human jaw (after Aeillo and Dean, 1990).

pressure on the incisor/canine region, but in hominids the muscle lies directly over the cheek teeth and acts vertically. In small-brained/large-jawed forms, including some early hominids, the area for muscle origin is extended by a sagittal crest, a bony flange running along the top of the skull. As brain size (and so vault surface area) increases, the entire temporal muscle of later hominids can originate from the side of the skull, and a crest is absent.

The second main muscle to raise the jaw is the masseter (the cheek muscle), originating on the zygomatic arch and passing backwards and downwards to the mandibular angle where corpus and ramus meet. A long zygoma will increase masseteric force through greater muscle size and efficiency by extending its origin further away from the mandibular pivot, so increasing the jaw lever's power arm; early hominids with a sagittal crest also show an anteriorly situated masseter as a feature of the mid-face.

The pterygoids originate on the pterygoid plates behind the palate and attach to the rear of the mandible: the lateral pterygoids in the condylar region, the medial pterygoids on the inner surface of the lower ramus. Acting together on both sides they pull the mandible forward; when the pterygoids on one side contract they pull over and partly rotate the condyle, so producing lateral movement of the tooth row. The combined action of temporals, masseters and pterygoids results in the vertical and lateral components of rotary chewing. In addition to these, there are also muscles in the mouth and throat, including the digastrics and mylohyoids forming the muscular floor to the mandible, that contribute to actions of food processing and vocalisation.

3.6 Transmission of chewing forces and mid-face structure

The forces required to process hard food items generate stresses in the face and jaws. The mandible needs to be strong enough to withstand such forces (see above), but they also pass upwards into the face. The zygomatic arches, acting as flying buttresses, transmit some forces laterally to the sides of the braincase, but the remainder pass up the face to the cranial vault. In large-jawed forms with projecting, faces, the maxillary region may be reinforced: the long canine roots act as supporting columns flanking the nasal aperture, and there may be a heavy buttress above the eyes (supraorbital torus), serving to tie face to braincase. Such an arrangement is seen in the African apes and in early hominids, where the reinforcement system may be elaborately developed (see chapter 5). In modern man these reinforcing structures have disappeared, partly because of the small weak nature of the jaws, and partly because of the unique vertical forehead of modern *H. sapiens* that enables forces to be transmitted directly to the frontal bone without the need for a heavy torus.

Face size and projection provide an obvious point of contrast between apes and humans. However, our reduced, flat (orthognathous) face is a recent evolutionary development, and early hominids had larger, prognathous faces, in keeping with their more powerful teeth and jaws. Moreover, there are differences in mid-face structure that cut across the hominid–pongid distinction and which have recently been recognised to have important phyletic implications (Ward and Kimbel, 1983). In the orang the premaxilla is strongly convex, curving back into the nasal aperture and slightly overlapping the anterior edge of the maxillary palatine process with which it forms the floor of the nasal cavity. The junction between the two is marked in the midline by a narrow incisive canal opening into the nasal and oral cavities by small incisive fossae. In the African apes the premaxilla is less convex and meets the anterior edge of the maxilla at a distinct step in the nasal floor, while the incisive canal is wide and the fossae broad and bowl shaped. Miocene hominoids from Asia resemble the orang (see chapter 4), while early hominids are like the African apes. The extreme facial retraction of modern humans and associated changes in premaxillary and maxillary ossification have somewhat modified the basic African ape–hominid pattern, but it is still evident in, for example, the large incisive canal.

3.7 Vocalisation

Speech sounds originate in the larynx, the upper part of the wind pipe just below where it opens with the oesophagus at the back of the throat. Here sounds are produced by expired air passing over the vocal cords (more accurately folds), causing them to vibrate. In other primates the larynx is well up the throat, close to the rear of the mouth and nasal cavity; in humans the larynx is lower down and there is a space, the pharynx, between the voice box and the oral cavity. Pharynx, mouth and nasal cavity together act as a resonator varied by movements of the tongue, mouth and soft palate, so altering the volume and pitch of sounds from the vocal cords and producing the different vowels. Consonants result from movements of the tongue, lips and teeth that momentarily interrupt the flow of expired air.

Constrasts in neck and throat anatomy between humans and other

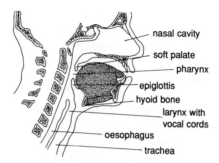

nasal cavity
soft palate
pharynx
epiglottis
hyoid bone
larynx with
vocal cords
oesophagus
trachea

Figure 3.4   Throat structures involved in vocalisation.

primates, especially the lack of a pharynx, mean that other species are physically unable to produce the range of sounds characteristic of human speech. Unfortunately, direct evidence for the evolution of the human vocal apparatus is meagre. There appears little or no correspondence between the soft structures responsible for speech and the adjacent skeletal anatomy, so that attempts to reconstruct the vocal apparatus of fossil hominids from details of the palate and basicranium remain controversial (see chapters 5 and 6). The brain's role in speech is summarised in section 3.8.

3.8   Brain

Brain development is considered characteristic of primates generally, but the great expansion and elaboration of the human brain is one of the most striking of evolutionary phenomena: brain size in humans averages $c. 1400 \, \text{cm}^3$ compared with $c. 400 \, \text{cm}^3$ for the chimpanzee and $c. 500 \, \text{cm}^3$ for the gorilla. This great increase is largely due to the unique human prolongation of the brain's fetal growth rate for the first year or so after birth, in contrast to all other primates, in which brain growth slows down after birth (Passingham, 1982; Martin, 1990). There is, however, very wide intraspecific variation in all hominoids (Tobias, 1971); the human range is $> 100\%$ (from $c. 1000 \, \text{cm}^3$ to $> 2000 \, \text{cm}^3$) without any obvious link to 'intelligence' or cognitive skills, however defined. At least part of this variability is size dependent, and in primate species with marked sexual dimorphism, such as the great apes, this is a major component of intraspecific diversity. Cross-species comparisons also need to take account of body size differences: brain size scales to body size with a ratio of < 1, so that simple brain to body weight ratios are inadequate (small animals have high ratios simply because they are small).

Regressions of brain size against body size over a range of species offer the most appropriate measures of relative brain size and provide a standard or 'expected' value for a given body weight against which actual sizes can then be compared. Several such formulae are available, differing in the range of species sampled and the formula judged to best describe the relationship between brain and body size. Two widely used ones are $E_E = 0.12W^{0.67}$ (Jerison, 1973) and $E_E = 0.05^{0.74}$ (Eisenberg, 1981; Martin, 1990), where E_E is the expected brain size and W is body weight. Both formulae are based on a wide range of mammalian species, and Jerison's in particular has been widely used, but different values provide a better fit for closely related species, e.g. the various hominoids (Holloway and Post, 1982; Martin, 1990). The ratio of actual brain size to expected brain size (E_A/E_E) is the encephalisation quotient (EQ). Holloway and Post (1982) show that the formula chosen can appreciably affect the EQ figures for hominoids, and that it is therefore most useful to express each species' EQ as a percentage of the modern human value. The formulae were derived from brain weights but it is usual to substitute cranial capacity instead, so allowing EQ calculations for fossils. Since the brain is surrounded by the protective meninges, endocranial volume exceeds brain size, but Martin (1990) shows it to be a reasonable approximation for these purposes.

The EQ figures show that the orang and gorilla are not especially encephalised, but chimpanzees have 2–3 times, and humans 6–8 times the brain size expected in a 'general' mammal of their body weights. While actual EQs differ between the two formulae, the proportional values are remarkably constant (Holloway and Post, 1982) and show that the great apes are c. 20–40% as encephalised as modern humans. New body weight estimates for fossil hominids show that the earliest species were, contrary to some earlier conclusions, hardly, if at all, more encephalised than the

Table 3.1 Brain size, body size, encephalisation quotient (EQ) and percentage EQ in modern hominoids. Data from Aeillo and Dean (1990).

	Brain size (g)	Body size (kg)	EQ Martin	EQ Jerison	%EQ Martin	%EQ Jerison
Pan troglodytes	410	36.4	2.38	3.01	38	37
Gorilla gorilla	506	126.5	1.14	1.61	18	20
Pongo pygmaeus	413	53	1.80	2.36	29	29
Homo sapiens	1250	44	6.28	8.07	100	100

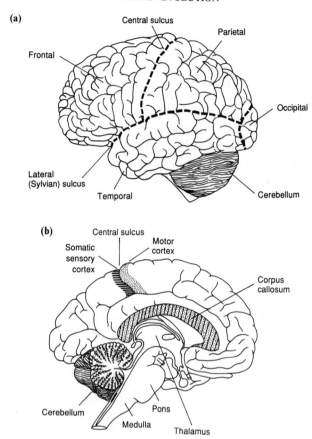

Figure 3.5 (a) Main features of the exterior of the brain (left hemisphere). (b) Section through the brain between the two hemispheres.

chimpanzee, but by the final Pliocene (c. 2.0–1.6 million years) brain expansion was under way (see chapters 5 and 6).

The main structures of the brain are shown in Figure 3.5. Expansion has most dramatically affected the cerebral hemispheres which surround and overlap the mid- and hindbrain structures such as the thalamus, pons, medulla and cerebellum. The pons and the medulla link the brain with the spinal cord and peripheral nerves, with the medulla also controlling autonomic functions such as breathing movements, heart beat and digestion. The thalamus sorts, correlates and interprets sensory impulses before transmitting them to the sensory cortex of the cerebrum. The cerebellum correlates complex muscular activities (conscious and unconscious) such as balance, movement and manipulation, and connects with the cerebral cortex through the thalamus and pons. The cerebellum is much enlarged in humans, reflecting the coordinating demands of truncal erectness, bipedalism and manual dexterity.

The outer layer of the cerebral hemispheres (the neocortex) consists of cell bodies ('grey matter'), its surface increased by being thrown into folds (gyri) separated by fissures (sulci) which also contain cortex, so that the total area of human neocortex is more than three times that of other primates after adjusting for body size (Passingham, 1975). However, brain evolution is not merely a question of size. There have certainly been structural and organisational changes to the 'circuitry', so that it is not possible to derive a human brain by 'adding bits on' to that of a chimpanzee or other primate. The very limited evidence available suggests that fundamental brain reorganisation occurred relatively early in hominid evolution and preceded the main increase in size (see below and chapters 5 and 6).

The central and lateral (Sylvian) sulci delimit the frontal, parietal and temporal lobes of each hemisphere. Additionally the occipital lobe may be delineated by the lunate sulcus, but this is often absent in humans although present in other primates. About 20% of the cerebral cortex is identified as primary sensory and motor areas, interconnecting with other structures such as the thalamus and cerebellum, and localised into body regions, with face, lips and tongue, and hands, fingers and thumbs especially well differentiated, reflecting the importance of facial expression in communication and manual dexterity and touch. Because nerve fibres cross over (decussate) before reaching the cerebrum, the left hemisphere controls the right side of the body and vice versa. Most individuals are right-handed, with neurological control of fine, precise manipulation located in the left hemisphere.

Most of the sensory and motor areas are on either side of the central sulcus, but the auditory cortex is in the temporal lobe and the visual cortex towards the rear of the cerebrum, confined to only a small area of the lateral side of the occipital lobe and running over the back and onto the medial surface. Non-human primates have a much larger area of visual cortex on the lateral side of the hemisphere, and its restriction in humans to the tip of the occipital lobe is indicative of the expanded association areas concerned with learning and integrative functions that make up c. 80% of the human cortex.

The two hemispheres are joined by the corpus callosum, allowing the transmission of impulses from one hemisphere to the appropriate region of the other, but there is clear evidence of specialisation between the two sides of the brain. The dominant hemisphere (almost invariably the left) controls speech, whereas the minor hemisphere is more concerned with visuo-spatial interpretation. Language control may be taken over by the right hemisphere following early childhood injury to the left one, but only at the expense of other cognitive skills, and by adulthood the capacity to switch hemispheres in this way is greatly reduced. Such specialisation would be advantageous as behavioural complexity increased (Passingham, 1982), as it would also be for intricate sequential manipulation. Hand preference appears a distinctively human trait, with most individuals right-handed (i.e. controlled by the left hemisphere). Passingham (1975; 1981) argues that handedness promotes complex manipulative skills by aiding learning (by reinforcement) of the preferred hand. If such skills or their contexts need to be articulated it would be a further advantage if they were controlled by the hemisphere also controlling language, hence their localisation in the left hemisphere of most individuals.

The hemispheres are anatomically asymmetrical, with the left generally larger than the right. Brain asymmetry is not confined to humans, but our pattern of left occipital and right frontal skew appears unique; the left hemisphere's specialisation for language contributes to this pattern.

Two areas of the cerebral cortex are particularly identified with language and vocalisation. Broca's area, towards the lower rear of the frontal lobe, controls speech fluency and articulation, whilst Wernicke's area, covering a larger area of the rear of the temporal and parietal lobes, controls speech comprehension. The two are connected by association fibres (the arcuate fasciculus). Wernicke's area lies adjacent to the auditory cortex (Heschl's gyrus) and is connected via the angular gyrus to the visual association area and visual cortex of the occipital lobe. Expansion of the left temporal lobe in the region of Wernicke's area contributes appreciably to the greater size of the left hemisphere in most individuals (Geschwind, 1972).

Speech involves the production of auditory patterns in Wernicke's area and their transmission to Broca's area, where articulatory patterns are produced which in turn are passed to the motor area immediately behind to stimulate movements of the mouth, throat, etc. 'Labelling' a seen object with its name involves the transfer of visual information from the visual cortex to the visual association area and then to the angular gyrus, which elicits the auditory pattern in Wernicke's area. Hearing the name involves the passage of signals from Heschl's gyrus to Wernicke's area and thence to the angular gyrus, which evokes the visual pattern to 'picture' the object in the visual association area.

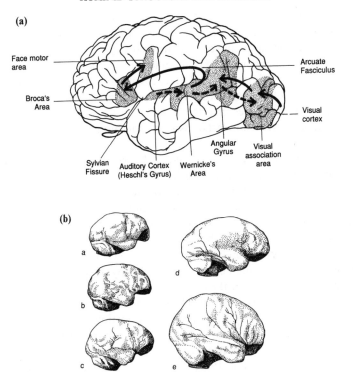

Figure 3.6 (a) The brain and language (after Geschwind, 1972) (see text). (b) Endocranial casts: (a) *Pan* (b) *A. africanus* (c) *A. robustus* (d) *H. erectus* (e) modern *H. sapiens* (after Holloway, 1974).

Language is closely and synergistically related to the development of consciousness and cognition: 'symbols devised for talking to others can as well be used for talking to oneself' (Passingham, 1982). These symbols (words) representing people, objects, events, sensations, etc. allow us to recall experiences from memory, rework and reformulate them, and so project and therefore plan ahead. This mental modelling promotes relational and other abstract concepts such as 'same', 'like', 'different', 'with', notions of sequencing and quantity, etc., and so further refines mental operations.

Associated with this is enhanced self-consciousness. Apes recognise themselves as distinct from other individuals, but only humans can provide a commentary on what they are experiencing or thinking. Self-monitoring of this kind ('my mind wandered; this is exciting; I'm bored') and imaginative accounts that involve the 'twisting' of reality into fiction and fantasy are aspects of human conceptualisation that are inextricably linked to language and appear to have no comparable parallels in other species although there is now evidence that some other primate species engage in deceit (Byrne and Whiten, 1988). Direct fossil evidence for these cognitive developments is, of course, lacking, but there is suggestive evidence from the archaeological record (see chapters 5–8).

Endocranial casts (endocasts) also provide limited insights into brain evolution. The cranial cavity reflects the brain's overall size and proportions, and faint impressions of sulci and gyri may be evident. However, these are generally indistinct, since the brain, cushioned by the meninges and cerebrospinal fluid, is not in direct contact with the vault bones so that cortical detail is usually poor, and information on subcortical structures is totally lacking. It is usually possible to locate the central and lateral sulci and delimit the primary lobes, but the lunate sulcus is commonly difficult or impossible to locate. Similarly, Broca's and Wernicke's areas cannot be identified as such, although clues are provided by the relative expansion of the appropriate regions of the cerebral lobes.

While early hominids had brains little or no larger than those of modern apes, they show contrasts in shape and proportions, and were probably organised differently (see chapter 5). Cerebral expansion was a relatively late phenomenon in human evolution (Holloway, 1983; Holloway and Kimbel, 1986; despite Falk, 1985; 1986), aspects of which are discussed in chapters 6–8.

3.9   Trunk

Whereas in quadrupeds the rib cage is narrow laterally and deep dorso-ventrally, in truncally erect forms it is expanded laterally and shallow from front to back, so keeping the centre of gravity over the hips. This shift of proportions alters the orientation of the clavicle (collar bone) and the position of the scapula (shoulder blade), which now lies on the back, not the side, of the rib cage, and the shoulder joint faces laterally rather than forward. The viscera, slung horizontally from the backbone in quadrupeds, are now supported from above, attaching to the underside of the diaphragm, and from below by the fleshy floor of the pelvic basin formed from modified caudal muscles that in quadrupeds move the base of the tail.

Support for the trunk is provided by the vertebral column, in humans made up of 26 vertebrae, each one capable of limited movement on its neighbours, so that overall the column has considerable flexibility. In erect hominoids the column is generally stouter than in quadrupeds, so aiding stability, and with the vertebrae increasing in size down the column since each carries successively more weight. The basal vertebrae in particular are massive and fused into a single mass, the sacrum, whose wedge shape aids stability since gravity forces it between the two hip bones.

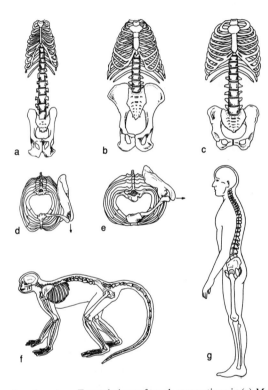

Figure 3.7   Trunk and posture. Frontal views of trunk proportions in (a) *Macaca* (b) *Pan* (c) modern *Homo*. Thorax and shoulder girdle of (d) *Macaca* and (e) modern human viewed from above. Arrows indicate orientation of shoulder joint. Posture and proportions of quadrupedal monkey (f) and erect human (g). In (f) the vertebral column acts as a spring under tension, with the alimentary tract a more or less horizontal tube slung below it. In (g) the vertebral column is more massively built towards its base for stability, and the basin shaped pelvis supports the viscera from below. The vertebral curves cancel out, so that weight passes directly to the lower limb. The iliofemoral ligament prevents the trunk rotating backwards at the hip joint while the cruciate ligaments lock the knee joint in place and allow the straight leg to bear the body's weight without muscle effort ((a)–(e) from Schultz, 1950; (f) from Le Gros Clark, 1962; (g) from Aeillo and Dean, 1990).

Above the sacrum the vertebrae are separated by spongy intervertebral discs, acting as shock absorbers and reducing friction. Their compression is largely responsible for the curves in the column, which enable it to withstand shocks and bend without breaking while still allowing efficient weight transmission; the curves cancel one another out, and weight passes from the head directly down to the hindlimbs. The newborn's spine has two curvatures, both directed forwards; additional curves develop as the infant lifts his head (secondary cervical curve) and then sits, stands and eventually walks upright (lumbar curve). This last, in particular, is an indicator of true truncal erectness and bipedalism.

3.10 Pelvis and hindlimb

The sacrum and hip bones (os coxae or innominate bones) form the basin-shaped pelvis. They are tightly bound by ligaments, so that while movement is limited it nonetheless allows absorption of jarring forces in locomotion and aids parturition in females. Each hip bone develops as three separate bones which fuse by puberty: the upper, fan-shaped ilium (the largest); the ischium at the base; and the pubis in front. The sacrum abuts against the ilia, and in humans the contact area is extensive, for each hip bone has to bear the entire body weight while the opposite leg is off the ground and swinging forward. For this reason the ilium is braced by two bony columns—the iliac pillar about one-third along from the anterior (ventral) edge of the bone, and a more extensive thickened area at its rear; both columns converge on the large, deep acetabulum (hip socket) and reinforce the hip bone.

Compared with other forms the human pelvis is shortened vertically and expanded laterally and dorso-ventrally so that the iliac blade is shorter, broader and more curved, and the ischium shorter and behind, rather than below, the hip joint. These changes minimise the distance for weight transmission between sacrum and lower limb, and also alter the action of muscles about the hip joint, so improving their mechanical advantage when moving the thigh and leg (see below). The line of weight transmission is between sacrum and hip joint; the tendency for the trunk to rotate backwards about the joint is prevented by the ilio-femoral ligament, while the forces tending to tear the sacrum away from the hip bone are resisted by the sacro-tuberous and sacro-spinous ligaments, and by the back muscles, which are under tension to hold the trunk upright.

From the trunk weight passes via the deep, tightly packed hip joint to the upper part of the leg. The femoral head is offset from the shaft by a distinct neck, internally braced to withstand the forces generated in bipedalism. The femora slope inwards, so that the expanded knee region lies directly below the hip joint. As the leg straightens the femur rotates medially and 'screws in' at the knee, close packing and so stabilising the joint, and tensing the collateral and anterior cruciate ligaments. When the leg is fully extended these lock the knee joint in position, and weight transmission is thus in a direct line down the straight limb. This allows support without muscular effort being required to counter body weight, and is a crucial adaptation for effective bipedalism.

The foot has to be stout enough to transmit body weight to the ground, strong enough to lift the body and sufficiently flexible to mould itself to surface irregularities, requirements resolved by the foot bones being formed into arches, two longitudinal (medial and lateral) and one transverse. This arrangement gives the foot the characteristics of a springy twisted plate rather than a flat flipper, and provides strength while retaining flexibility and resilience, without which the foot would shatter. The arches are

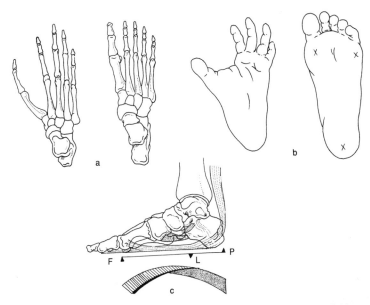

Figure 3.8 (a) Foot skeletons of chimpanzee (left) and human (right). (b) Feet of chimpanzee (left) and human (right), with points of ground contact marked by crosses. (c) The foot as a lever and a twisted plate. F = fulcrum; L = load (body weight); P = power provided by calf muscles. (see text) ((a) after Schultz, 1963; (b) after Campbell, 1966; (c) after Aeillo and Dean, 1990).

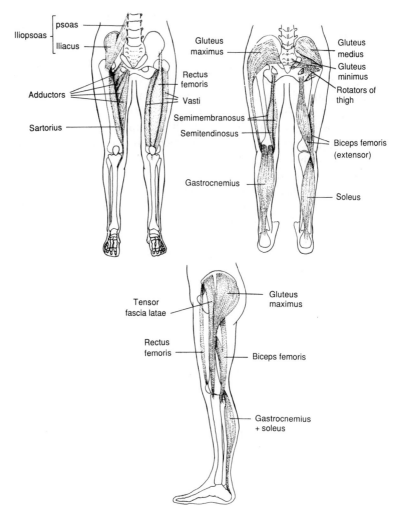

Figure 3.9   Front, rear and lateral views of some of the main muscles involved in modern bipedalism.

maintained by ligaments and the plantar aponeurosis—a band of fibrous tissue running along the underside of the foot and acting as a bowstring to keep the longitudinal arches under tension. Weight is directed to the ground at the base of the big and little toes and at the heel, a tripod-like pattern aiding stability, while the arches make the foot a flexible but strong and stable lever, aided by the prominent, backwardly projecting heel (calcaneum) which increases the length and efficiency of its power arm, with the effort provided by the calf muscles (gastrocnemius and soleus).

3.11 Bipedalism

The above summary deals with weight transmission in static terms; movement involves additional forces which the locomotor system must be able to withstand. Human bipedalism is a curious, even bizarre, locomotor activity involving several constituent actions that make for flexibility. Although other species (primate and non-primate) engage in bipedalism of various kinds, none approaches humans in their set of energetically efficient features. The convergent femora and locking knee joint noted above are especially important here, but there are other adaptations as well. Striding, which is unique to humans, involves alternate stance and swing phases of each leg: if the right leg is in stance phase (i.e. ground contact) the trunk is pulled over to the right side by muscle action and the pelvis tilting, while also rotating around the right hip joint so that the left leg swings forward, pushed off by the toes. The left leg then strikes the ground at the heel with the knee slightly bent to absorb the shock, so beginning its stance phase. The pelvis tilts over to the left while muscles pull the trunk over to that side, and the pelvis rotates about the left hip joint; the right foot, meanwhile, is entering swing phase by pushing off at the ball of the foot behind the big toe, aided by a lateral twist of the calcaneum (heel bone), which wedges itself against the cuboid immediately in front and transforms the foot into a rigid lever.

This sequence is produced by several groups of muscles and involves the following actions: flexion and extension of the thigh on the hip; pelvic tilt; pelvic rotation; adduction; flexion and extension of the leg on the thigh; flexion, extension and rotation of the foot on the leg. The shape of the hip bone greatly aids efficient muscular action (see above). The main flexors about the hip joint are muscles running down the front of the thigh: rectus femoris, arising from the anterior inferior iliac spine and inserting on the knee cap; iliacus (from the iliac blade's inner surface) and psoas (from the lumbar vertebrae). These converge on the lesser trochanter and upper part of the femoral shaft, and also pull the trunk laterally during stance phase.

The main extensors of the thigh are the hamstring muscles—biceps femoris semimembranosus and semitendinosus—arising from the ischial tuberosity in a mechanically advantageous position behind rather than below the acetabulum, and inserting on the top of the tibia and fibula; by their contraction they straighten the thigh, providing the main power in walking. They can be reinforced by gluteus maximus, the large, fleshy buttock muscle originating on the back of the ilium and sacrum, but the primary role of this muscle appears to be postural, maintaining the trunk erect. The other smaller gluteal muscles, medius and minimus, originating on the side of the ilium and inserting on the greater trochanter of the femur, are important in pulling the trunk over during pelvic tilt, so aiding locomotor efficiency and preventing collapse towards the unsupported side of the body.

Movements of the leg below the knee are principally flexion and extension. The main flexor of the leg on the thigh is biceps femoris and the main extensor quadriceps femoris—a group of four muscles, including rectus femoris (see above). The remainder are the vasti, passing down the front of the thigh to insert around the knee, and by their contraction straightening the entire lower limb.

The main extensors of the foot are the calf muscles gastrocnemius and soleus, inserting via the Achilles tendon on the back of the heel. By their contraction the heel is raised and the body lifted off the ground, as with standing on tiptoe (both legs) or in the propulsive 'toe-off' of bipedal striding (each one alternately). The foot flexors are, by contrast, poorly developed in man. There are also muscles that rotate and twist the foot and, on the underside, muscles that reinforce the ligaments and plantar aponeurosis in maintaining the arches during locomotion, as well as toe muscles similar to those of the fingers (see below).

The upshot is that human *walking* is energetically rather more efficient than most forms of quadrupedalism at low speeds, and appreciably more so than chimpanzee bipedalism or knuckle walking (Rodman and McHenry, 1980). As speed increases efficiency falls off, so that running is energetically expensive, suggesting that bipedalism evolved in response to selection pressures for efficient low-speed terrestrial locomotion, perhaps in sparse environments (chapters 5 and 6).

3.12   Forelimb

The human forelimb has the same basic structure as the hindlimb, but lacks its weight-bearing adaptations and is capable of wider movement. Some features of the shoulder girdle are largely a consequence of truncal erectness (see above), and there are similarities here with other species such as the African apes. However, these also show adaptations to weight bearing in their forelimbs, a consequence of their knuckle-walking locomotion, that are lacking in hominids. The hand is, in fact, the most distinctive part of the human forelimb, and its range of finely controlled movements, ranging from a delicate, gentle touch to a grip of great pressure, forms the basis for the acquisition and elaboration of technology.

The upper arm bone (humerus) articulates with the scapula (shoulder blade) at the shallow, open shoulder joint. The scapula rides over the back

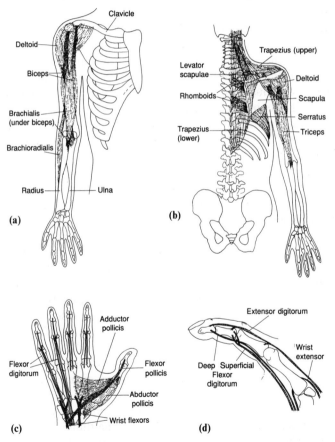

Figure 3.10   Shoulder girdle and forelimb. (a), (b) Main muscles involved in movements of shoulder, arm and forearm. (c) Main muscles of the thumb and fingers. (d) The hand and digits as jointed levers showing the insertions of flexor and extensor muscles.

of the rib cage, held by muscles and braced by its joint with the clavicle (collar bone). Muscles rotate the scapula on the rib cage, so altering the orientation of the shoulder joint and permitting a wide range of arm movements. The muscles include trapezius (upper fibres) and serratus, which rotate the scapula so that the shoulder joint faces upwards, and trapezius (lower fibres), levator scapulae and rhomboids that rotate the scapula so the joint faces downwards.

The deltoid muscle, originating on the clavicle and scapula, inserts on the humerus and abducts and raises the arm at the shoulder. Flexion of the forearm at the elbow results from contraction of biceps (originating on the scapula), brachialis and brachio-radialis, originating on the humerus. The first two attach to the forearm (radius and ulna) near the elbow, the last to the radius near the wrist. Extension of the elbow is effected by triceps, originating on the scapula and humerus and inserting on the olecranon process of the ulna behind the elbow. Biceps also act to rotate the forearm by pulling the radius over the ulna so that they lie parallel and the palm faces forwards (supination) when the arm hangs down by the side of the torso. Contrary rotation, where the radius is pulled over and across the ulna so that the palm faces backwards (pronation), is brought about by pronator muscles of the forearm.

The hand and its five digits are essentially a series of jointed levers with movements at various points controlled by complex neural pathways. Rotation of the forearm results in pronation (palm upwards) and supination (palm downwards) movements, while the eight, small, irregular bones of the carpus (wrist) allow considerable mobility of the hand (contrast the weight-bearing ankle and foot). In the palm are the metacarpals moving on the carpus with muscles producing flexion and extension at the wrist, and beyond these the phalanges of the individual digits—two for the thumb and three for each finger. Since the basal phalanx can move on its metacarpal (at the knuckle), the middle phalanx on the basal, and the distal phalanx in the middle, the digits also form a series of segmented levers. Each digit has flexors (on the palm side) and extensors (on the back of the hand); particularly important for manipulative skills are the strong flexors and extensors of the index finger (flexor and extensor indicus) and of the thumb (flexor and extensor pollicis).

In addition the digits can be splayed out by appropriate abductors and drawn together again by adductors, those of the thumb being especially significant. The trapezium at the thumb's base has a saddle-shaped surface upon which the metacarpal rides, allowing rotation and the wide range of thumb movements characteristic of the human hand. Adductor pollicis,

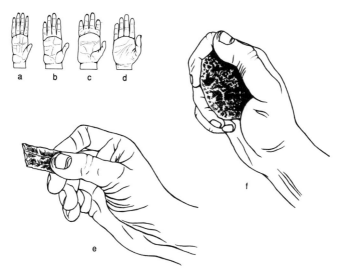

Figure 3.11   Palm and finger proportions of (a) orang, (b) chimpanzee, (c) gorilla, (d) human (after Aeillo and Dean, 1990). (e) and (f) Human precision and power grips (modified from Napier).

forming the fleshy mound at the base of the thumb, draws it across the palm and towards the other digits, leading to pulp to pulp contact, especially with the index finger. The resulting precision grip is capable of great power and at the same time is delicate and finely controlled. It distinguishes hominids from other primates capable only of a power grip (clenching the fist and crossing the thumb over the bunched fingers) or, in some cases, a weak precision grip, in which the relatively short thumb can be adpressed against the lower part of the index finger but pulp to pulp contact is not possible.

Archaeological evidence indicates that all tool kit assemblages, even the earliest, contain at least some tool types that require an essentially human precision grip for their manufacture and/or use (see chapters 5 and 6).

CHAPTER FOUR

THE CATARRHINE RADIATION AND HOMINID ORIGINS

Hominid evolution is merely one aspect, albeit for many the most fascinating, of the wider primate diversification, itself part of the mammalian adaptive radiation. Hominids, together with the Old World apes and monkeys, form a major primate group, the catarrhines. These, together with the New World monkeys and tarsiers, comprise a diurnally adapted, predominantly plant-feeding suborder of primates primarily reliant on vision for sensory data, the Haplorhini. In contrast, for the nocturnal, partly or wholly insectivorous lemurs and lorises (Strepsirhini) olfaction is a major sense.

Hominids are currently represented by only a single world-wide species (*Homo sapiens*), although fossil evidence indicates an earlier, more extensive degree of species diversity. The extant apes, restricted in range and few in numbers, are relicts of a much more extensive radiation in the mid-Tertiary, whereas the Old World monkeys (Cercopithecoidea) comprise a large number of highly adaptable and successful species widely distributed over much of Africa and Asia. Their expansion occurred relatively recently, so pointing to differential timing for the radiation and diversification of the Old World higher primates.

The living apes are widely regarded as the most human-like and therefore most 'advanced' of other primates, and this has been translated into the supposition that they are the primates phyletically most closely related to ourselves. Evidence from both fossil and extant forms indicates that this is only partly correct, and 'apes' are now generally recognised as a diverse group. It is usual to distinguish at least between the lesser apes or Hylobatidae (gibbons, *Hylobates;* and siamangs, *Symphalangus*) of South-East Asia and the greater apes—chimpanzees (*Pan*) and gorilla (*Gorilla*) of Africa and the orang-utan (*Pongo*) of Borneo and Sumatra, often placed together in the primate family Pongidae. It is now evident that this last group, including all the most 'man-like' apes, is itself phyletically diverse.

The three African ape species, the common chimpanzee (*Pan troglodytes*), pygmy chimpanzee or bonobo (*Pan paniscus*) and gorilla (*Gorilla gorilla*, sometimes included in *Pan* as *P. gorilla*), are clearly closely related. They have many cranial and dental features in common and share distinctive modifications of the shoulder, elbow, wrist, hand and digits that enable the forelimb to withstand the compressive forces associated with their terrestrial knuckle walking. Their growth patterns are remarkably similar, and in numerous features (but not all) they can be considered scaled variants of each other, so that many of their distinguishing characteristics are simple consequences of differing body sizes.

The orang differs from the African apes in many cranio-dental features (Andrews and Cronin, 1982) and lacks their knuckle-walking complex, possessing instead unique fore- and hind-limb adaptations for arboreal clambering. Biochemical, chromosomal and DNA data all support this assessment, based on traditional morphological evidence, of the African apes' close phyletic relatedness and the distinctiveness of the orang.

In fact, over the last two decades it has become increasingly clear that man, not the orang, is the African apes' closest relative. This conclusion is surprising given the profound morphological contrasts between apes and humans, but the evidence, largely from biochemical and genetic studies, is compelling. Studies of a range of hominoid proteins including serum albumins, carbonic anhydrase, myoglobin, haemoglobin, cytochrome C and fibrinopeptides, together with hybridisation and sequencing studies of nuclear and mitochondrial DNA, have all provided data on hominoid phylogeny. Useful reviews are King and Wilson (1975), Cronin (1983), Goodman *et al.* (1983), Sibley and Ahlquist (1984), Andrews (1986a,b), Jeffreys (1989) and Martin (1990).

Findings naturally differ in their details but point towards one major conclusion: African apes and humans share so many details of their protein and DNA sequences in which they contrast with the orang that they must have shared a common ancestry well after the orang lineage split off. Indeed, chimpanzees and humans are estimated to have *c*. 98.5% identity of the bulk of their nuclear DNA (Jeffreys, 1989), while their protein molecules are similar to those between closely related (sibling) species, and appreciably more alike than those of many congeneric species of other groups (King and Wilson, 1975). Nor is the evidence exclusively nonmorphological: there are features of the upper and mid-face, cheek, palate and dentition that are shared by African apes and humans to the exclusion of the orang (Andrews and Cronin, 1982).

In other words, the family Pongidae as commonly understood is paraphyletic (see chapter 2) since the sister group of African apes (ourselves) has been removed to a separate family, Hominidae. Some of the latter's distinctive features, such as reduced canines and bipedal locomotion, evolved early as basal hominid adaptations, while others (increased brain size, facial reduction) were later developments. Overall, however, the hominid clade has been characterised by rapid morphological evolution compared with the African ape lineages.

Recognition of 'ape' paraphyly, of the need to distinguish between primitive, derived and independently acquired characters, and that both hominid and ape radiations were previously more extensive than present species diversity and distribution suggest has powerfully influenced the conceptual framework of palaeoanthropology. There is now greater awareness of the possibility of multiple species in the past (with the corollary that only a small proportion of known fossils can be on, or even close to, the ancestry of living forms), and so less predilection to force all available fossils on to the lineages leading to one or other of the few extant hominoids.

In fact it has become increasingly clear that primate evolution has included multiple radiations, some wholly extinct, and that overall primate diversity is more extensive than recognised only a few years ago. Space limitations here preclude all but the briefest of summaries strongly biased towards the immediate context of hominid origins. Fortunately there are several recent accounts that provide excellent overviews of primate evolutionary biology and phylogeny. See, for example, Conroy (1990), Fleagle (1988) and Martin (1990). These authors differ in their interpretations, reflecting the uncertainties and complexities of primate evolution. Primates appear relatively early (70–65 million years ago (mya)) in the placental fossil record and are initially known mainly from North America and, to a lesser extent, Europe. The earliest forms (paromomyids, plesiadapids) are bizarre, with large jaws bearing pronounced, rodent-like anterior teeth, and claws rather than typical primate nails. There is now some doubt whether they should be recognised as primates at all (Martin, 1991). At the very least they are distinguished as plesiadapiformes or 'archaic primates' from euprimates or 'primates of modern aspect'.

These more recognisably primate forms had already begun to radiate by the early Eocene, some 54 mya; the initial expansion was of strepsirhines — small-brained forms reliant on olfaction as a primary sense with simple placentation, probably mainly nocturnal and wholly or partly insectivorous. Modern strepsirhines include the lemurs, lorises and galagoes, but these are bigger brained than the Eocene forms, and show numerous other contrasts. By 50 + mya early haplorhines had appeared. These were diur-nally active forms, reliant on vision rather than smell as the major sense, with complex placentation and a more elaborate central nervous system; in general appearance they most closely resembled modern *Tarsius*, although lacking some of its specialisations, such as those of the locomotor system.

By the final Eocene (40 + mya) there are remains of larger haplorhines with deeper faces and shorter, stronger mandibles containing vertically implanted anterior teeth and large, more elaborate molars. These allowed more effective processing of vegetation and represent forms that approxi-mate to a monkey grade and niche. The earliest remains (*Pondaungia* and *Amphipithecus*) are from Burma, with somewhat later (38–33 mya) and more diverse forms from Oligocene deposits of the Fayum, Egypt. In general appearance and niche these probably resembled some of the smaller extant monkeys but were markedly more primitive, and there are no direct links to any living species. Indeed, at this period the term 'monkey' is meaningless in a phyletic as opposed to a grade sense, and the Fayum forms are best thought of as stem catarrhines, i.e. basal Old World 'higher' primates.

Apart from a fossil tarsier (*Afrotarsius*) and a specimen of uncertain status (*Oligopithecus*), two main groups, Parapithecidae and Propliopithecidae, are usually recognised within the Fayum fossils. Parapithecids are the more numerous; they retain some primitive features (such as three premolars) but also show dental specialisations that make it unlikely that they are closely related to any modern primates. Propliopithecids are rather larger but less numerous; they possessed a typical catarrhine dental formula of 2.1.2.3, having lost the anterior premolar, and they lack specialisations that would exclude them from the ancestry of later catarrhines. On the other hand, they show many primitive features and cannot convincingly be closely linked to any later forms.

Abundant plant remains indicate that the Fayum environment was probably forest swamp, and dental morphology suggests most species were frugivorous, an inference also consistent with their estimated body sizes. Several fossils are sufficiently complete to indicate size contrasts, plausibly interpreted as sexual dimorphism, in at least some species. This is characte-ristic of many catarrhine primates and is associated with inter-male rivalry for mating access to females, suggesting a complex polygynous social organis-ation like that of many monkeys.

The earliest ceboids (New World monkeys) date from the late Oligocene (*c.* 25 mya). Opinion differs as to whether the Old and New World simians represent separate, parallel developments from distinct antecedents, or whether the New World monkeys evolved from Old World forms that reached South America by rafting. The former view is the traditional one, but in recent years the latter has gained support as a result of greater knowledge of continental drift and the undoubted similarities, cranial and postcranial, between some of the Fayum forms and modern ceboids.

The Fayum deposits sample an early phase of catarrhine evolution; evidence of a much expanded radiation is provided by numerous Miocene fossils (24–5.5 mya) with their wide geographic range (Szalay and Delson, 1979; Conroy, 1990). The first group are from lower- to mid-Miocene (24–16 mya) East African sites around Lake Victoria and in north Kenya. They represent some six or more genera and ten to a dozen species grouped as the Proconsulidae. Some forms are known only from jaw fragments and teeth, others by incomplete cranial material, but one species, *Proconsul africanus*, is represented by more extensive remains including an almost complete skeleton. They are linked by common dental features: broad upper incisors; tall, narrow lower incisors; quadrate upper molars and long lower molars with thin enamel. All evidence indicates that proconsulids were cranially and postcranially more advanced than the Fayum forms but still primitive compared with modern catarrhines.

Proconsulids varied considerably in size, ranging from that of a small monkey (*Micropithecus*) to a large chimp (*Proconsul major* and *Afropithecus*), and in facial proportions and postcranial morphology. The contrasts doubtless reflect locomotor and feeding differences, with the larger forms at least partly terrestrial. The finds are associated with a variety of environments (forest, woodland, bush) and some species were evidently sympatric.

The best-known species, *P. africanus*, had a lightly constructed cranium, together with unequal, rather short and stoutly built limbs that made it fairly monkey-like overall, but with some ape-like features such as the absence of a tail. At around 10–12 kg it was similar to a medium-sized monkey and was probably an arboreal quadruped, lacking the suspensory specialisations of modern apes, although another genus (*Dendropithecus*) shows rather more suspensory features in its forelimb. Recent discoveries at Kalodirr, north Kenya (Leakey and Leakey, 1986a,b), extend the proconsulid range. *Simiolius* and *Turkanopithecus* are small and medium forms, while *Afropithecus* is similar to *P. major* in size. It is distinctive, however, in its long narrow snout, projecting incisors and apparently thicker dental enamel.

Proconsul was long considered ancestral to the chimpanzee and gorilla, but its similarities with the living African apes, which are mainly dental, are probably plesiomorphies and there are no convincing indicators of close phyletic links. Palatal morphology is similarly primitive: the premaxilla is well separated from the bony palate by a broad, deep incisive fossa rather than the narrow incisive canal of modern apes. Indeed, it is now clear that the group is best viewed as broad adaptive array diversified to exploit distinct niches, rather than constrained into the ancestry of any modern ape.

More or less contemporary with the proconsulids are jaw fragments and isolated teeth of *Victoriapithecus*, which show the molar ridge development (bilophodonty) for processing leaves characteristic of cercopithecoids, indicating the separation of the Old World monkey clade. Whilst there is nothing to clearly link the *Proconsul* group to any extant ape, *Victoriapithecus* does at least indicate separation of cercopithecoid and non-cercopithecoid (hominoid) catarrhine groups by the early/middle Miocene.

The relatively warm, moist environments of the period saw expansion of primary forest over much of Europe and Asia, and numerous primate species exploited this zone. The fossils are later than the proconsulid radiation (mid- to late Miocene, 16–10 mya) and the larger European forms are generally referred to as *Dryopithecus* (oak ape) because the original finds were discovered with fossilised oak leaves. The specimens show somewhat thicker enamel than the *Proconsul* group, narrower, less spatulate incisors and molars with low, rather rounded cusps suggesting frugivory; the upper molars tend to be narrow compared with the quadrate *Proconsul* teeth. Two species, *D. fontani* and *D. laietanus*, are usually recognised, the former better known than the latter. Isolated postcranial remains assigned to *Dryopithecus* suggest a locomotor system more like that of modern hominoids than *Proconsul*, but without the knuckle-walking specialisations of the modern African apes.

Other European hominoids include *Pliopithecus*, a medium-sized form once considered a gibbon ancestor but now recognised as remarkably primitive in many of its features and overall broadly resembling some New World monkeys with suspensory adaptations. *Oreopithecus* is a larger, probably arboreal, form displaying forelimb elongation and some truncal erectness. It shows a puzzling combination of characters, including several dental specialisations that remove it from close phyletic relationships with other forms. Some workers have argued for the origins of both *Pliopithecus* and *Oreopithecus* among the early East African proconsulid array, but this is far from certain.

Broadly contemporary with *Dryopithecus*, and extending from central and south-east Europe through the Middle East to the Indian subcontinent and China, and with finds as far south as Kenya, are fossils of the *Sivapithecus/Ramapithecus* group. Their systematics are particularly confused, with some workers recognising several genera and others recognising only one, *Sivapithecus*. Sivapithecids were larger than most earlier apes (about male orang/female gorilla size) and more dentally derived, with low, robust canines, and larger, low-crowned, closely packed cheek teeth with rounded cusps and thickened enamel. These were originally thought to be hominid traits, and in the 1960s to 1970s *Ramapithecus* was widely regarded as the base of the hominid clade, which was thought to have differentiated by at least *c*. 14 mya. This interpretation collapsed in the late 1970s as more complete specimens from Turkey and Pakistan showed *Sivapithecus* to resemble closely the orang in facial and dental features (see below). Moreover, further fossils eroded the difference between *Ramapithecus* and *Sivapithecus* so that many workers now include the former genus within the latter as *S. sivalensis*. These discoveries, and the detailed similarities between modern African apes and humans, rule out *Ramapithecus* as a basal hominid, and the dental resemblances probably reflect the functional requirements for processing a hard diet of seeds, nuts and hard-pitted fruits. Most fossils are derived from contexts appreciably drier than those of earlier hominoids, indicating a mosaic open/mixed woodland environment. Most *Sivapithecus/Ramapithecus* remains are 14–10 my old with the youngest fossils *c*. 8 my old.

European specimens include an early form, *S. darwini* from Czechoslovakia (14 my old) and enigmatic material from Rudabanya, Hungary, originally assigned to two distinct genera (*Bodvapithecus* and *Rudapithecus*) but which almost certainly represents species of either *Sivapithecus* or *Dryopithecus*. There are also upper and lower jaws and a partial skull from Greek sites dated to *c*. 11–10 mya assigned to *Ouranopithecus* (*Graecopithecus*) (de Bonis *et al.*, 1990; Andrews, 1990) that differs from other sivapithecid fossils in several important respects (see below).

Asian sivapithecids are comparatively well known, particularly from discoveries in the Siwalik Hills, Pakistan, and from Lufeng, China. The Siwalik sample dates from the later Miocene (10–8 mya) and includes jaws, crania and postcranial remains of two species: *S. indicus* and the smaller *S. sivalensis* (including *Ramapithecus*). A partial skull of *S. indicus* displays high orbits, weak brow ridges, flaring zygomatic arches, narrow nose and snout with a projecting premaxillary region as in the orang (Pilbeam *et al.*, 1982). A similar pattern is seen in the rather earlier (12–10 mya) partial face of *S. meteai* from Turkey (Andrews and Tekkaya, 1980). Siwalik postcranial

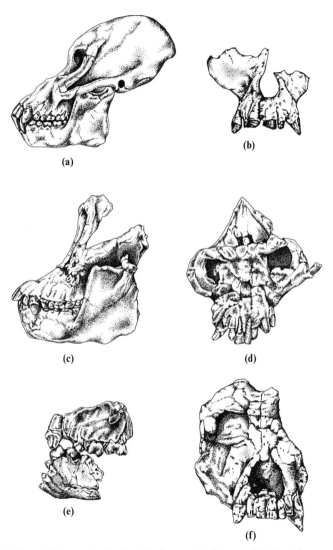

Figure 4.1 Some Miocene hominoids. (a) *Proconsul africanus*, (b) *Sivapithecus meatei*, (c) *Sivapithecus indicus*, (d) *Lufengpithecus*, (e) *Kenyapithecus wickeri* and (f) *Ouranopithecus* ((b) from Andrews and Tekkaya, 1980; (c) from Ward and Pilbeam, 1983; (d) from Conroy, 1990; (f) from de Bonis, 1990).

remains include arm, hand, thigh and foot bones indicative of arboreal clambering and climbing, although the mosaic contexts suggest some ground activity as well.

There are also numerous remains from Lufeng, Yunnan, Central China. Distinct from the Siwalik fossils in their broader upper face and other features, the Lufeng crania again show resemblances to orangs. Originally referred to *Sivapithecus* and *Ramapithecus* (Oxnard,1987), it is now argued that the material represents male and female of a single species, *Lufengpithecus lufengensis* (Conroy, 1990), appreciably more sexually dimorphic than any living hominoid (Kelley and Xu, 1991; Martin, 1991). Closely related to, and probably derived from, *Sivapithecus* is the appropriately named *Gigantopithecus*, characterised by increased body size and expanded cheek teeth together with robust, low canines that in older individuals are worn flat to the level of the tooth row. An early almost gorilla-sized species, *G. giganteus* (= *G. bilaspurensis*) is known from final Miocene Siwalik deposits and a later form (*G. blackii*), perhaps twice gorilla size, from Plio-Pleistocene deposits in China. Size and dental morphology indicate a terrestrial folivorous niche, perhaps exploiting bamboo forest.

Sivapithecid remains from Africa were poorly known until recently, but the sample has been greatly extended by discoveries near Lake Victoria and in north Kenya (Pickford, 1985a, b). These fossils date from the mid-Miocene (14–12 mya) and so are earlier than the Siwalik and Chinese fossils. Most are referred to *Kenyapithecus africanus* (called by some *S. africanus*) with a second species (*K. wickeri*) based upon limited material from Fort Ternan, Kenya. While thickly enamelled and dentally rather like the Asian fossils, upper jaw material indicates that *K. africanus* retains the plesiomorphic *Proconsul*-like pattern of the nasal floor, with a broad incisive fossa separating premaxilla and palate.

Cranio-dental similarities (including thick enamel) between sivapithecids and the orang point to an Asian pongid clade including these forms and *Gigantopithecus*. However, the presence of sivapithecids in Europe and Africa suggests caution, and the possibility that some shared features typify mid- to late Miocene hominoids in general, so that some aspects of orang morphology may be better thought of as relict rather than distinctively Asiatic.

However, at least one trait appears a reliable phyletic indicator. Asian sivapithecids (Turkey, Siwalik, Lufeng) and the orang share a highly derived condition of the nasal floor, with the strongly arched premaxilla overlapping the palate and only a narrow incisive canal (Ward and

Pilbeam, 1983). *Proconsul, Dryopithecus, Kenyapithecus* and *Ouranopithecus* share the apparently more primitive pattern of a broad incisive fossa separating premaxilla and palate, suggesting that Asian sivapithecids (and the orang) are derived from forms approximating the Afro-European specimens. The fossils' chronology accords with this interpretation.

In other words, it looks as though the sivapithecid radiation originated in the mid-Miocene, perhaps from within the proconsulid group, and subsequently split into two clades: an Afro-European one perhaps ancestral to the present chimp and gorilla (and humans), and an Asian one, which subsequently differentiated to include lineages leading to the recently extinct *Gigantopithecus* and still extant orang.

That said, the differentiation of the present African ape lineages is obscure since the African fossil record almost totally peters out after Fort Ternan and other mid-Miocene sites. There are a few fragmentary remains and isolated teeth covering the period 12–4 mya (see below), but otherwise nothing is known of chimpanzee and gorilla ancestry. Were it not for the living species, we would be forced to conclude that African apes dwindled and became extinct by the final Miocene.

The issue has been investigated another way by using biochemical and genetic data to construct a molecular clock for primate evolution on the assumption that changes in the amino acid and DNA sequences are selectively neutral and accumulate at a constant rate. It is difficult to accept that functional proteins can be strictly neutral as components of varied phenotypes over extended periods in differing habitats, but neutrality is perhaps more likely for DNA sequences, since much of the molecule is non-coding 'junk'. This still does not ensure a constant substitution rate and, indeed, there is growing evidence of rate variation in different lineages (Jeffreys, 1989). Martin (1990) provides a valuable review of molecular clocks in primate evolution.

Differences between contemporary species form the clock's basis, and an initial speciation event (date of separation of two lineages) is needed to calibrate it; once set it is assumed to keep good time, and divergence dates for other groups can be estimated in proportion to the split used for the initial calibration. Andrews (1986a,b) provides hominoid differentiation dates based on fossil evidence for the origin of the *Sivapithecus*–orang lineage at 13–14 mya. There is clear evidence for an African ape–hominid clade after the orang's separation, and agreement on the close relationship and recent split of the common and pygmy chimpanzees, but no clear pattern of chimpanzee, gorilla and hominid divergence emerges. Their recency of common ancestry and molecular similarities may be beyond the resolving powers of currently available techniques, especially when the complexities of back-mutation and convergence are borne in mind.

Some biochemical and genetic studies point to the chimpanzee as more closely related to us than it is to the gorilla, while others indicate chimpanzee and gorilla to be more closely related to each other than either is to humans, thereby supporting the traditional view of their affinities. Another possibility is a trifurcation, i.e. virtually simultaneous speciation events leading to the differentiation of chimpanzee, gorilla and hominid lineages. Examples of molecular clock dates include estimates of 8–10 mya for gorilla divergence and 6.3–7.7 mya for the chimpanzee–hominid split based on DNA hybridisation techniques, and a possible trifurcation at 6.3–8.1 mya based on nuclear DNA sequencing (Andrews, 1986b).

All these approaches have their difficulties: chimp and gorilla morphological similarities point to a common ancestry after the African ape–hominid split, in which case protein and DNA similarities between chimp and humans must be convergent features. If chimps and hominids share a common ancestry after the differentiation of the gorilla lineage the same must apply to chimp–gorilla similarities, both molecular and morphological. For example, either knuckle-walking adaptations evolved independently in chimp and gorilla (improbable) or knuckle walking also characterised the hominid–chimpanzee common ancestor and was subsequently lost in one descendant lineage (hominids) but retained in the chimpanzee clade.

While some workers (e.g. Zihlman and Lowenstein, 1983; Zihlman, 1989) have argued for a knuckle-walking pygmy chimpanzee-like ancestor for hominids, many see major difficulties with this interpretation. Knuckle walking and bipedalism are alternative locomotor modes, and it is difficult to envisage forms behaviourally and morphologically committed to knuckle walking making the switch to effective bipedalism. While negatives cannot be proved, the complete absence of any knuckle-walking features in even the earliest hominid fossils lends support to this view.

Despite the lack of ape remains, other catarrhines are well represented in the Mio-Pliocene record: fragmentary remains of possible hominids are present in the Rift Valley from c. 6 mya onwards, and are increasingly frequent after 4 mya, while numerous cercopithecoid fossils are known from late Miocene and Pliocene deposits, indicating that several important monkey groups (e.g. baboons, geladas and colobines) were radiating during this period. It is surely significant that the initial appearance and early evolution of hominids, and the more widespread and diverse radiation of Old World monkeys, occurred during the period that saw the contraction and virtual disappearance of apes. The same ecological factors were almost certainly involved: reduced rainfall and increased seasonality leading to contraction of forest and expansion of grassland. The extent of this shift has sometimes been overstated, and evidence indicates a mosaic of forest, woodland and open environments in East Africa, for example (Hill, 1987), but there is no doubt that in many regions the extensive forests of the mid-Miocene were replaced by much patchier, discontinuous tree cover with more open areas of bush or grassland.

Early hominids, in this context best regarded as derived apes, were able to effect the shift from forest cover to more open environments, as were the

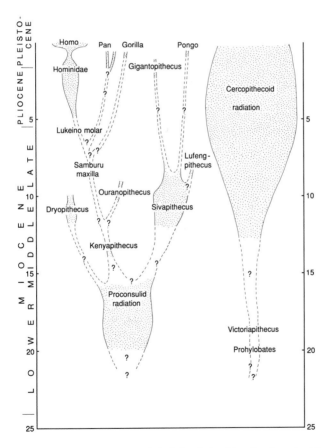

Figure 4.2   Possible adaptive diversity and phyletic relationships of major catarrhine groups. The scheme is much simplified and, in parts, highly speculative.

cercopithecoids. The latter group, with their folivorous adaptations, were also better able than the frugivorous apes to exploit effectively the food resources of dwindling tree-covered areas. Despite the early occurrence of *Victoriapithecus* (see above), Old World monkey fossils only became numerous from the final Miocene, increasing in frequency and diversity during the Pliocene. Semiterrestrial, open-country forms such as geladas, baboons and their extinct relatives are especially well represented.

The same ecological changes underlying the marked reduction in ape numbers and diversity evidently provided opportunities for the adaptive radiation of the cercopithecoids, whose success further contributed to ape decline. Indeed, the burgeoning monkey radiation may have been a significant factor in the very origin of hominids.

In addition to the primary catalysts of climatic change and ensuing floral and faunal shifts, another, more local, factor was probably involved in hominid origins. The earliest possible hominids are from various East African sites where the developing Rift system of the Miocene and Pliocene altered topography and climate, creating a rain shadow and changing drainage patterns and lake levels, with further effects on vegetation and faunal communities. The extant African apes are confined to forest and woodland west of the Rift, whereas fossil and living monkeys and early hominids are known from the drier savannah and grassland areas of the East African Rift system and the South African high veld.

Details of early hominid evolution are still largely unknown, but the clade probably originated in the later Miocene or basal Pliocene, perhaps as an ape species adapted to drier, more open environments with discontinuous tree cover, but evidence to test this suggestion is lacking. Several East African sites have yielded fragmentary remains from this period (5–14 mya). The earliest specimens, a talus from the Muruyar Beds (> 13 mya) and teeth from the Ngorora formation (10–13 mya), of the Tugan Hills sequence, near Lake Baringo, are only slightly younger than forms such as the Fort Ternan *Kenyapithecus* and perhaps represent late species of the *Proconsul* radiation, or descendant forms, as may a younger maxilla from the Namurunguk formation, Samburu Hills (8 mya). A lower molar from the Lukeino formation (5.5–6 mya) of the Tugen Hills shows some chimpanzee-like features and may represent a stage just prior to, or immediately post-dating, the hominid–African ape split (Hill and Ward, 1988).

Rather later specimens (< 5.5 mya) are more clearly hominid, and are dealt with in the next chapter. By the time more complete evidence is available (< 3.5 mya) several critical adaptive complexes—masticatory system and dentition, trunk and hindlimb locomotion—had already evolved characteristically hominid patterns, and their possessors were clearly emancipated from a forest niche. The search for adequate fossil evidence in the period 4–10 mya to document the origin and initial evolution of hominids and the context, mode and tempo of such critical adaptations as rotary chewing, truncal erectness and bipedal locomotion remains one of the great challenges for palaeoanthropology.

CHAPTER FIVE

PLIOCENE HOMINIDS

5.1 Introduction

Definite hominid fossils are known from Pliocene and basal Pleistocene sites (c. 5.5–?1.4 mya) in South and East Africa. In the earlier part of the period (> 4 mya) finds are few and fragmentary and confined to East Africa, but they turn up in increasing frequency after about 3.5 mya, and significant numbers are known from later Pliocene localities and around the Plio-Pleistocene boundary. They are usually referred to the genus *Australopithecus* and include most major skeletal parts so that the morphology, size and proportions of these early hominids are known quite well.

The first finds were from cave in-fills in South Africa, but over the last 30 years many discoveries have been made in the East African Rift system. They include spectacular finds at Olduvai Gorge and Laetoli (Tanzania), Koobi Fora and west Turkana (Kenya) and Hadar (Ethiopia), which yielded, among others, 'Lucy'—probably the best known of all early hominids. *Australopithecus* was evidently widespread, long lived ($\geqslant 4$ my) and morphologically diverse, with several species recognised. Its main features and phyletic and adaptive relationships are summarised below. Later specimens extend to the Plio-Pleistocene boundary and beyond (see chapter 6), when they overlap with early members of the genus *Homo*. The possible ecological and behavioural interactions of these forms, together with details of the sites that reveal sympatry, are further discussed in chapter 6.

5.2 Historical

In 1925 Dart described a fossilised child's cranium from Taung near Kimberley, South Africa (Dart, 1925). Because of its recovery so far south of the chimpanzee and gorilla ranges, Dart named the specimen *Aus-*

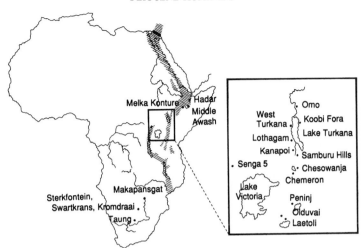

Figure 5.1 Some important Pliocene and basal Pleistocene hominid sites. Hatching indicates the Rift Valley system.

tralopithecus africanus ('southern ape of Africa') while recognising its hominid status. The size and proportions of the brain (revealed by a natural endocast), the wear pattern, morphology and proportions of the teeth and the shape of the dental arcade all pointed to human, not ape, affinities, and the cranial base and position of the foramen magnum suggested truncal erectness. At the time few accepted Dart's interpretation, and the fossil was generally considered to be an aberrant extinct ape.

However, in the 1930s and 1940s, Broom, excavating limestone caverns near Krugersdorp, Transvaal, recovered further hominid fossils. At Sterkfontein he found cranial, dental and postcranial material which he assigned to *Plesianthropus transvaalensis* ('near man of the transvaal') (Broom, 1937; Broom and Schepers, 1946), now included in Dart's species *Australopithecus africanus*. The Sterkfontein site continues to be excavated, and has yielded the largest sample of *A. africanus*. At Kromdraai, less than a mile away, Broom recovered the partial skull of a larger, more rugged individual that he named *Paranthropus robustus* ('robust beside man'), (Broom, 1938), and at Swartkrans, also nearby, more extensive remains of similarly large forms with expanded cheek teeth were identified as *Paranthropus crassidens* ('great toothed beside man') (Broom, 1949; Broom and Robinson, 1952). The Kromdraai and Swartkrans fossils are generally lumped together within the species *A. robustus*, although recent years have seen a growing tendency to distinguish between them (see below). *Paranthropus* is now sunk by many within *Australopithecus*, although again there has been a recent trend influenced by cladistic studies (see chapter 2) to distinguish between the two genera.

In the late 1940s Dart returned to excavation, and with Tobias and Hughes recovered further remains of *A. africanus* (originally *A. prometheus*) from Makapansgat, some 200 miles north of Johannesburg. By the late 1950s sufficient fossil material was available to indicate two hominid morphologies: a smaller, more lightly built 'gracile' form (*A. africanus*) and a larger, heavier form with expanded cheek teeth (*A. robustus*). Discoveries over the last 30 years have confirmed this picture in some respects and significantly modified it in others. The South African specimens have

Table 5.1   Some important early hominid sites.

Site	Age (mya)	Fossils	Species
Lothagam, Kenya	5.5	Mandible fragment	*Australopithecus* sp.
Tabarin, Kenya	4.5	Mandible fragment	*Australopithecus* sp.
Kanapoi, Kenya	4.0	Humerus	*Australopithecus* sp.
Middle Awash, Ethiopia			
Belohdelie	> 4.0	Cranial fragments	*A. afarensis*
Maka	3.5–4.0	Femur	*A. afarensis*
Hadar, Ethiopia	2.8–3.4	Crania, mandibles,	*A. afarensis*
Omo, Ethiopia	1.8—3.1	Cranial fragments, mandibles, postcrania, teeth	*A. afarensis*, *A. aethiopicus*, *A. boisei, H. habilis*
Lomekwi, Kenya (West Turkana)	2.5	Cranium	*A. aethiopicus/A. boisei*
Other West Turkana sites	1.8–3.4	Crania, mandibles postcrania	?*A. afarensis*, *A. aethiopicus/A. boisei*
Laetoli; Tanzania	3.6	Cranial fragments, mandibles, postcrania	*A. afarensis*
Makapansgat, South Africa			
Members 3, 4	2.5–3.0	Crania, mandibles postcrania	*A. africanus*
Sterkfontein, South Africa			
Member 4	2.5–3.0	Crania, mandibles postcrania	*A. africanus*
Taung, South Africa	?2.0 +	Cranium	*A. africanus*
Swartkrans, South Africa			
Member 1	?1.5–1.8	Crania, mandibles, postcrania	*A. robustus*
Kromdraai, South Africa	?2.0	Cranium, mandible, postcranial	*A. robustus*

continued to accumulate, but since the late 1950s discoveries have extended the range of *Australopithecus* to East Africa.

In 1959 Louis and Mary Leakey recovered from Olduvai Gorge, Tanzania, the cranium of an especially robust australopithecine, which they named *Zinjanthropus boisei*, now recognised as *Australopithecus boisei* (Tobias, 1967). Further discoveries from Peninj, Tanzania, the upper levels of the Shungura Formation in the Omo Valley, Ethiopia, and from Baringo and Koobi Fora, Kenya, have confirmed the morphology of a hyperrobust East African australopithecine species with massive cheek teeth. Virtually all these specimens are well dated at around 1.5–2.0 mya, although the discovery of a specimen (KNM WT 17 000) at Lomekwi, West Turkana (Walker *et al.*, 1986), suggests *A. boisei* or something very like it may date back as far as 2.6 mya (see below).

From 1972 Mary Leakey shifted her excavation to Laetoli, some 30 miles south of Olduvai, recovering a number of hominid palatal and mandibular remains dated to *c.* 3.6 mya. Also during the 1970s Johanson, Coppens and Taieb, excavating at sites along the Awash river at Hadar, Ethiopia, dated between 3 and 3.4 mya, recovered highly variable dental, cranial and post cranial hominid remains. The finds were extensive: one Afar locality (AL 333) yielded remains of at least 13 individuals, another (AL 288) the partial skeleton of a small female ('Lucy'). Johanson *et al.* (1979) combined all this material with that from Laetoli to form a single, highly variable species, *A. afarensis*. Individual teeth from the lowest levels of the Shungura and Usno Formations, Omo (2.9–3.1 mya), have also been assigned to *A. afarensis*, as have some of the fragmentary remains from earlier Rift Valley sites such as the Lothogam mandible and Kanapoi humerus (see chapter 4).

There is therefore evidence from both South and East Africa of a broadly similar sequence: earlier, cranially lighter hominids (*A. africanus*/*A. afarensis*) in the mid-upper Pliocene, and later, more ruggedly constructed, megadont forms in the final Pliocene/basal Pleistocene (*A. robustus*/*crassidens*, *A. boisei*), but with regional contrasts between the approximately contemporary hominid forms (see below).

5.3 Australopithecine species

Most remains of *Australopithecus* consist of skulls and teeth, so that the genus and its species have been recognised and largely defined on cranial and dental features; postcranial remains are fewer and generally less

variable. The cranial contrasts are mainly in face, jaw and dental proportions, and are usually considered to reflect shifts in feeding and diet.

Compared with later hominids all australopithecines share a common
pattern of a small, unexpanded braincase (400–550 cm³, i.e. within the
chimpanzee and gorilla range) combined with a large face and jaws. To this
extent they resemble apes, but there are many important features that
confirm hominid status. While jaws range from large to massive and the
mandible is often heavily buttressed, the dentition is hominid in shape and
proportions. The tooth row is evenly curved and in most individuals the
incisors are unexpanded and canines non-projecting; dental enamel is thick
or very thick. Despite this, tooth wear is extensive but the pattern indicates
rotary chewing, not vertical chomping. The projecting face is strongly
constructed and the zygomatic arch long, with masticatory power mainly
directed over the cheek teeth, which in all australopithecines are large (and
in the robust forms massive) compared with other primates.

Despite the small brain size, cerebral height is greater than in apes, and
the frontal and temporal lobes are more human-like, with likely expansion
of Broca's area, and left hemisphere dominance (Holloway, 1972b; 1974).
The cranial base is relatively long in earlier fossils, more flexed in the later
robusts, but the foramen magnum and occipital condyles—the junction
with the vertebral column—are set relatively far forward. The neck muscles
attach low down on the back of the skull and there is a prominent mastoid
process, suggesting effective head balance on an erect trunk. This is
confirmed by postcranial fossils which indicate truncal erectness and
bipedal locomotion, although there is debate as to the pattern and efficiency

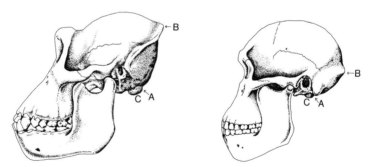

Figure 5.2 Comparison of pongid (*Gorilla*) left, and early hominid (*Australopithecus
africanus*) right. Note contrasts in relationships of face and braincase, vault proportions, size
and morphology of anterior dentition, position and angle of contact with vertebral column (A
and C) and extent of nuchal musculature on rear of vault (B) (from Le Gros Clark, 1962).

of australopithecine bipedalism (see below). Contrary to earlier impressions, recent analyses (Jungers, 1988b; McHenry, 1988) suggest that *interspecific* size differences were minor, but that all species showed marked internal variability in body size, probably reflecting sexual dimorphism (see below).

The main cranial and dental features of the different species are as follows.

5.3.1   A. africanus

This, the first recognised species, is best known from the large collection from Sterkfontein member 4, with a smaller sample from Makapansgat and the original, immature specimen from Taung. This last site has been quarried away, but the geology of the Sterkfontein and Makapansgat caverns has been investigated in detail. Six members have been recognised within the breccia in-fill at the large Sterkfontein site. All except the lowest (member 1) are fossiliferous, but hominid remains do not occur until member 4, with its abundant *A. africanus* fossils. On the basis of comparisons with East African faunas, Vrba (1982) estimated the member 4 fauna to be *c*. 2.8–2.3 my old and the overlying member 5 to be *c*. 1.8–1.5 my old. Delson (1988) gives a broadly similar age for member 4 of 2.5 my based on cercopithecoids. Sedimentology and faunal data indicate increasing seasonality in member 4 and a marked shift towards greater aridity between members 4 and 5 (see below). It has recently been suggested that the member 4 breccia may span a considerable period of time (Vrba, 1988) and that significant evolution may have occurred within the populations represented by the fossil hominids (Kimbel and White, 1988).

At Makapansgat the breccias are divided into five members with *A. africanus* remains from member 3, and just one specimen from the overlying member 4. Faunal comparisons suggest an age of 3.0–2.6 my for member 3, and this is broadly compatible with palaeomagnetic data from the site. There is evidence of two short reversed intervals near the base of member 4 within a longer period of normal polarity. It has been suggested that these correspond to the Mammoth and Kaena events (3.1 and 2.9 mya respectively) within the Gauss normal epoch (see pp. 25–27), indicating an age of $\geqslant 3.1$ my for member 3 and up to 3.1 my for member 4. Overall, the evidence suggests that Makapansgat may be slightly older than and/or partly overlap with Sterkfontein.

The more complete *A. africanus* crania show a common pattern of a lightly constructed, domed braincase with a capacity of *c*. 400–500 cm^3 and moderate to large jaws and dentition. The palate is long and the dental arcade

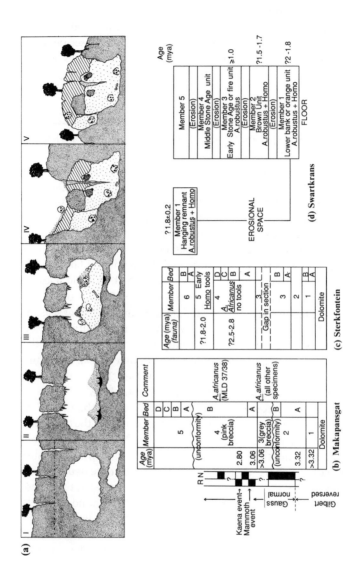

Figure 5.3 South African hominid sites. (a) Schematic sequence of cavern formation: I, acidic ground water enlarges faults and fissures in limestone to form underground caverns; II, water table falls and rain water enlarges the fissures until surface contact occurs, allowing air and fine debris into cavern; III, clefts enlarge, and larger particles and animal remains fall or are washed into cavern. Percolating water consolidates debris into hard breccia; IV, weight of breccia causes cavern floor to collapse into chamber below, disturbing stratigraphy. Later cavern roof collapses or erodes away, allowing breccia to accumulate faster. Water channels carry recent material down into deeper levels of cavern deposit; V, softer limestone erodes away, leaving harder breccia plug as small hillock. (b) Makapansgat sequence and chronology. (c) Sterkfontein sequence and chronology. (d) Swartkrans sequence and chronology ((a), (d) after Brain, 1981, 1988; (b), (c) after Tobias, 1981).

(b) Makapansgat

Age (mya)	Member	Bed	Comment
	5	D	
		C	
		B	*A.africanus*
		A	(MLD 37/38)
	(unconformity)		
2.80	4 (pink breccia)	B	*A.africanus* (all other specimens)
>3.06		A	
3.06	3 (grey breccia)	B	
	(unconformity)		
	2	A	
3.32			
>3.32	Dolomite	1	

(c) Sterkfontein

Age (mya) (fauna)	Member	Bed	
	6	B	Early Homo tools
?1.8–2.0	5	A	
	4	D	
?2.5–2.8		C	*A. africanus* no tools
		B	
	3	A	
	Gap in section		
	3	B	
	2	A	
	1	B	
	Dolomite	A	

(d) Swartkrans

	Age (mya)
Member 5	
(Erosion)	
Member 4 Middle Stone Age unit	
(Erosion)	
Member 3 Early Stone Age or fire unit *A.robustus*	≥1.0
(Erosion)	
Member 2 Brown Unit *A.robustus* + Homo	?1.5–1.7
(Erosion)	
Member 1 Lower bank or orange unit *A.robustus* + Homo	?2–1.8
FLOOR	

Member 1 Hanging remnant *A.robustus* + Homo	?1.8±0.2

EROSIONAL SPACE

narrow compared with later hominids, with the front teeth set in a curve, so that there is moderate to extensive prognathism. Incisors are generally moderate in size and canines non-projecting, but in some individuals (e.g. Stw 252) incisors are large and canines project significantly beyond the occlusal plane. The mandible is variably developed with some specimens relatively lightly built but others more strongly constructed, with a reinforced symphysis and thickened corpus.

Other evidence points to a powerful masticatory apparatus: the cheek teeth are relatively and absolutely large compared with other catarrhines, while the zygomatic arch is long, and provides an extensive origin for the masseter muscle. The area of temporalis origin is also extensive, with a well-defined perimeter (especially anteriorly) and, judging from incomplete material, some individuals possessed a sagittal crest to increase the area of attachment.

The middle face is relatively flat with depressed nasal bones, while the nasal aperture is bounded laterally by bony swellings over the canine roots. A brow ridge, weak above the orbits but more strongly developed in the midline, buttresses the upper face to the braincase. The cranial base is variable, being relatively long and unflexed in some individuals and more flexed in others, but in all specimens the nuchal musculature is only moderately extensive.

5.3.2 A. robustus

South African 'robusts' were first recognised on recovery of the Kromdraai skull, but are much better known through the many specimens from Swartkrans. These are often crushed or otherwise damaged through deposit collapse within the cavern's complex stratigraphy (Brain, 1981; 1988). Member 1, the oldest unit, is represented by two distinct entities: an orange basal deposit and a 'hanging remnant' at the back of the cave; the intervening mass was apparently swept away by water erosion. Erosional intervals also separate member 1 from the brown member 2 breccia, and this from the overlying members 3 and 4. *A. robustus* fossils and tools are known from members 1–3 and early *Homo* from members 1 and 2. Faunal comparisons suggest member 1 to be 2.0–1.8 my old, member 2 c. 1.5 my old and member 3 around 1 my old. The fauna also indicates a dry open environment in member 1, but with some patches of tree cover.

The nearby Kromdraai site, now little more than a narrow cleft on a small hill, represents the remains of a large cavern, most of which has been eroded away. Of its five breccia members, the few hominid specimens are

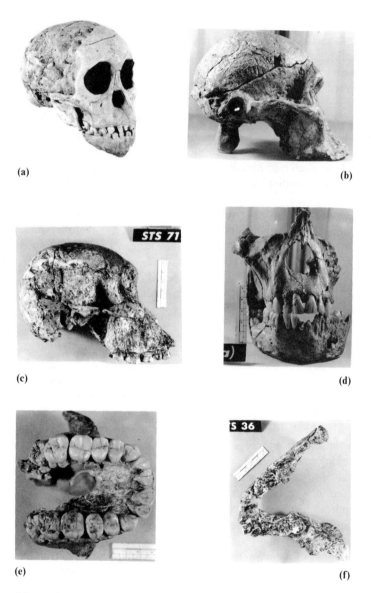

(a)

(b)

STS 71

(c)

(d)

(e)

S 36

(f)

Figure 5.4 *A. africanus.* (a) Taung. (b) Sts 5. (c) Sts 71. (d) Sts 52 a and b. (e) Sts 52a occlusal. (f) Sts 36. Note dental wear.

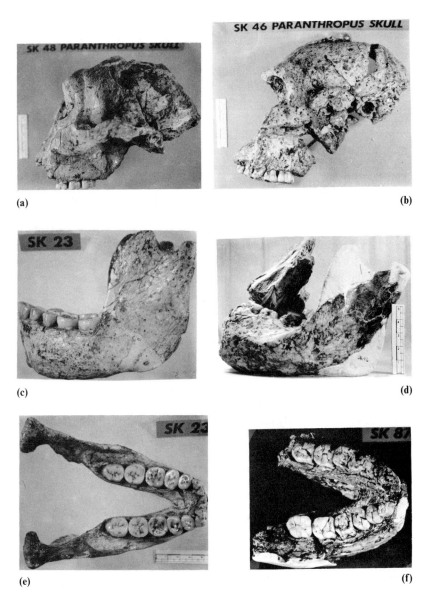

Figure 5.5 *A. robustus.* (a) SK 48. (b) SK 46. (c) SK 23. (d) SK 12. (e) SK 23 occlusal. (f) SK 876 occlusal. Note expanded cheek teeth and crowded anterior dentition.

from member 3, dated on faunal and palaeomagnetic grounds to 2.1 mya. The environment was rather wetter than Sterkfontein 4 or Swartkrans 1, with more tree cover.

In this species the braincase is associated with a larger and heavier jaw apparatus and dentition than *A. africanus*, with a modified facial architecture. The palate is long, the mandible large to massive and constructed accordingly. Among the most distinctive features of many robust specimens is the greater development of the cheek teeth and the relatively (not absolutely) small size of canines and incisors, set in a straight line across the front of the jaw so that there is little of the premaxillary prognathism found in *A. africanus*, although the alveolar border is expanded in the cheek region. These characteristics are more developed in the Swartkrans sample than in the Kromdraai specimen, and are among the features cited by those who distinguish between *A. crassidens* and *A. robustus* from the two localities.

The dental features affect facial proportions: the profile is concave, the zygomatic arch powerfully built, long and laterally flaring with deep maxillary processes, the anterior faces of which slope downwards and forwards to increase the area for masseter attachment. In profile they obscure the nasal region so that the entire middle face is set in a central hollow bounded by the projecting maxillary processes and overhung by the heavily built, handlebar-shaped supraorbital torus. Bony pillars flanking the nasal region converge medially below the torus, further reinforcing the facial architecture. These features represent masticatory adaptions in the robust australopithecines, both to provide a more extensive origin and greater mechanical efficiency for the jaw muscles and to resist stresses generated during chewing; the area of *temporalis* is extensive, and a distinct sagittal crest occurs in mature crania.

The braincase is lightly built, although provided with crests and tori, and is proportioned differently from that of *A. africanus*. It is somewhat larger ($500 + cm^3$) and set at a lower level relative to the face, so that it does not dome as in *A. africanus*. There is greater lateral expansion, especially in the mid-vault, and increased flexion of the cranial base, possibly associated with the enlarged cheek teeth and longer palate and/or the more anterior positioning of the expanded cerebellum under the cerebrum.

5.3.3   A. boisei

Robust East African crania assigned to *A. boisei* display most of the features of *A. robustus* to a greater degree. The cheek teeth, especially the premolars, are noticeably larger, and so total area for chewing is greater. Since the anterior teeth in *A. boisei* are no larger, and in some cases smaller, than in the South African species, contrasts between anterior and posterior dental proportions are even more marked. The larger *A. boisei* crania are bigger than any Transvaal specimens, and bear more pronounced crests and tori. The zygomatic arch is long and widely flaring and its maxillary process deeper; so far forward does the process extend that the bony pillars running up over the canine roots and prominent in *A. robustus* (and to a lesser extent in *A. africanus*) are in *A . boisei* totally obscured by a bony flange formed by the maxillary zygomatic process (Rak, 1983). The supraorbital torus is strongly built, especially medially, and the mandible massive, with reinforced buttressing both at the symphysis and running along the body of the jaw below the cheek teeth. The mandibular ramus is high so that the occlusal plane is set well below the jaw pivot, increasing the antero-posterior component to occlusion (and so grinding efficiency), especially in the premolar–molar region.

In addition to massive *A. boisei* crania, often considered to be those of males, a number of smaller specimens (e.g. KNM ER 732—see chapter 6) have been cited as females. If this is correct the species shows a greater range of cranial variation than *A. robustus* with both larger (male) and smaller (female) crania than in South Africa. However, alternative specific attribution for KNM ER 732 is possible (see chapter 6).

The above summary applies quite well to almost all larger *A. boisei* specimens. However, the recently recovered cranium (KNM WT 17 000— the 'black skull') from Lomekwi, west Turkana (Walker *et al.*, 1986) extends the age of robusts back to 2.5 mya and presents several distinctive features: despite its great size the specimen has only a modest cranial capacity (400 cm^3) and well-developed sagittal and nuchal crests. The cranial base is long and flat, rather than flexed as in later *A. boisei*. The cheek teeth were large judging by their roots (the crowns are missing) but so were the anterior teeth, set in a distinct curve in the massive, prognathous upper jaw. The cranium's morphology combines overall features of *A. boisei* with similarities to *A. afarensis* and has important phyletic and systematic implications (see below). Because of the contrasts with later *A. boisei* specimens it is considered by some to represent a new species, *A. aethiopicus*, named from specimens first recovered from the lower Omo (see below), whilst others are content to include it within *A. boisei*. Irrespective of name, all are agreed that it takes the origins of the East African robust clade back to 2.5 + mya.

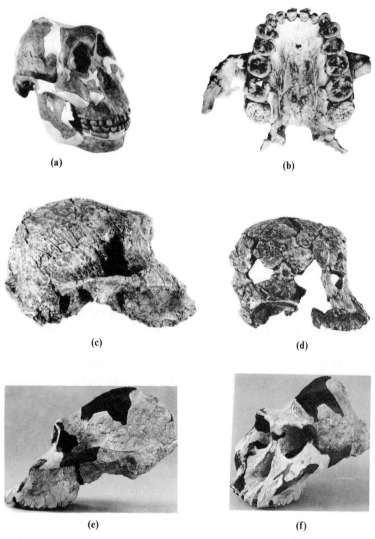

Figure 5.6 *A. boisei and A. aethiopicus.* (a) reconstruction of OH 5. (b) OH 5 palate. Note expanded cheek teeth and small anterior teeth. (c) KNM ER 406. (d) KNM ER 732. (e, f) two views of KNM WT 17 000.

5.3.4 A. afarensis

In 1978 Johanson *et al.* recognised a new East African australopithecine species, *A. afarensis*, on the basis of specimens from Laetoli and the much larger and highly variable samples from Hadar, Ethiopia (see above). The Laetoli material is well dated at between 3.59 and 3.77 mya while the Hadar specimens are from three members separated by volcanic tuffs: the Sidi Harkoma (SH) member, the oldest; Denan Dora (DD), yielding most material; and Kadar Hadar (KH), containing some specimens, including 'Lucy'. Initial potassium–argon dates (Aronson *et al.*, 1977) gave an age of 2.8–3.3 mya with most specimens from the younger end of that range; further estimates, considered more reliable, give an age of 3.65 ± 0.15 my for within the SH member (Walter and Aranson, 1982), suggesting that the hominid specimens cover a period of 2.9–4.0 mya. However, Brown (1982) provides geochemical data indicating identity between the SH tuff at the base of the Hadar sequence and the Tulu Bor tuff at Koobi Fora dated at 3.35 mya so supporting the initial, younger, estimate and suggesting that the bulk of the Hadar fossils are *c.* 3.4 my old or less.

The Laetoli specimens consist largely of jaw fragments and teeth (White, 1977); those from Hadar include a much wider range of skeletal elements from numerous individuals including a partial skeleton AL 288A–'Lucy'. The most useful reviews are those by Johanson and White (1979), White *et al.* (1981) and Johanson *et al.* (1982b). The hominid specimens are highly variable so that early publications suggested three Hadar species, gracile and robust australopithecines and early *Homo*, although from the late 1970s Johanson *et al.* considered that only a single, highly variable, species was represented.

Johanson *et al.* (1982) also regard the Laetoli and Hadar specimens as similar enough to warrant pooling into a single taxon, which in their view differed significantly from other australopithecines. The main contrasts emphasised by these workers are cranial: larger anterior teeth (broad spatulate incisors and asymmetric projecting canine crowns); smaller cheek teeth; a narrow, shallow palate with tooth rows which converge posteriorly; a convex and projecting premaxillary region with prominent juga over the canine roots forming lateral pillars to the face but not extending as far upwards as in *A. africanus*, whilst the zygomatic arch is lightly built and relatively short, arising from the maxilla above the region of P^4/M^1 or M^1 and separated from the canine pillar by a deep fossa. The cranial vault has a capacity similar to that of *A. africanus* but is marked by pronounced mastoids which project laterally, compound temporal/nuchal

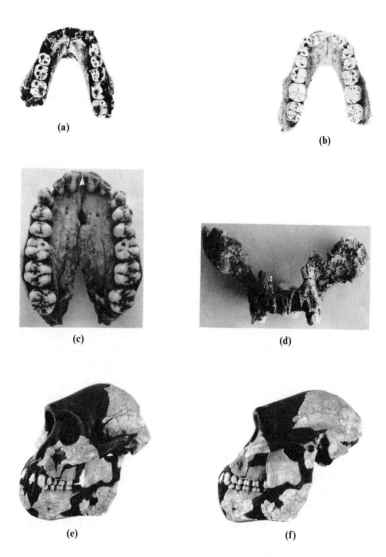

Figure 5.7 *A. afarensis.* (a) Laetoli LH4 mandible. (b) Hadar AL 400-1A mandible. (c) Hadar AL 200 1A palate. (d) AL 333-1 facial fragment. (e,f) two views of composite skull reconstruction by Kimbel *et al.*

crests and closely adjacent temporal lines running along the vertex of the vault. The Hadar and Laetoli mandibles have straight rather than curved postcanine tooth rows and unexpanded mandibular bodies.

White *et al.* (1981) interpret this morphology as reflecting the large anterior teeth of *A. afarensis* with their robust, curved roots and with little masticatory emphasis on the posterior dentition. They consider it a more primitive morphology than that of *A. africanus* and they regard *A. afarensis* as the stem hominid, ancestral to both later australopithecines and to the genus *Homo*.

A composite reconstruction of the *A. afarensis* skull (Kimbel *et al.*, 1984) emphasised its primitive nature with prognathous face, large, spatulate incisors, overlapping canines and retreating frontal and flattened parietals, so that the overall impression is remarkably chimpanzee-like. A later, revised reconstruction (Kimbel, *et al.*, 1988) modifies this with a shorter, higher vault and more vertically expanded frontal. A recently recognised frontal fragment (AL 33–125) from Hadar shows that at least some *A. afarensis* individuals possessed a well-developed sagittal crest, with emphasis on both anterior and posterior temporal muscle fibres (Asfaw, 1987).

A. afarensis has attracted criticism on several grounds:

(1) Its initial announcement in 1978 did not strictly conform to the rules of the International Code of Zoological Nomenclature, so making for subsequent taxonomic difficulties. This is a 'legalistic' rather than biologically substantive objection, but conformity with the code is important for ensuring clarity and avoiding ambiguity in taxonomy.

(2) The type specimen eventually designated is from Laetoli (LH 4), whereas the bulk of the material is from Hadar. Early comparisons with other species were made on the basis of the combined sample, asserting but not demonstrating that the material from the two sites was sufficiently similar to be pooled into a single species.

(3) At least some of the cranial and dental features cited in the original diagnosis as characteristic of *A. afarensis* are demonstrably present in *A. africanus* (Tobias, 1981). The recently recovered Sterkfontein Stw 252 specimen (Clarke, 1988) with its large incisors and projecting canines further supports character overlap.

(4) Johanson and White's phyletic interpretation of *A. afarensis* as the stem hominid ancestral to both later australopithicene species and *Homo* has been much criticised.

More detailed descriptions and further data have helped clarify some of these issues. The publication of separate dental data for Laetoli and Hadar

(White, 1980; 1985) reveals them to be sufficiently similar to be pooled; while there is undoubtedly overlap in many cranial features between *A. afarensis* and *A. africanus*, there are certain features in the Laetoli and Hadar samples that are not present in the South African fossils. Olson (1981; 1985) argued on cranial evidence that more than one species was represented within the Hadar sample, but this was refuted by Kimbel *et al.* (1985; Kimbel and White, 1988), who argued that cranial variation is within the range expected from modern and fossil comparators. This interpretation has apparently since then been accepted by Olson (Grine, 1988a). Overall the case for a single, although highly variable, species on the basis of the cranial and dental evidence seems pretty well founded, although the postcranial remains are more contentious in this respect (see below).

Also regarded as *A. afarensis* are the Belohdelie vault fragments and the Maka femur, from adjacent localities in the Middle Awash, some 40 miles south of Hadar. A prominent marker at both sites is the cindery tuff (CT) dated to 3.9–4.0 mya. The Belohdelie fragments include much of a frontal bone from a level below the CT and so are > 4.0 my old (White, 1984). The frontal bone is short and broad, with only slight postorbital constriction compared with other australopithecine frontals, and remarkably flat (Asfaw, 1987). The Maka fossil is the upper end of a subadult femur showing structural and morphological adaptations to bipedalism and which closely matches one of the larger Hadar femora in size. It was recovered from above the CT, and faunal comparisons suggest an age of 3.5–4.0 my—as old as, or older than, the Laetoli footprints.

Very early and fragmentary rift hominids are also sometimes included with *A. afarensis* (Hill and Ward, 1988). They include the Lothagam and Tabarin mandibles (5.5 and *c.* 4.5 my old) and the Kanapoi humerus (*c.* 4 my old). Inclusion is generally on similarity in size and proportions with Hadar fossils, but can only be tentative because of the specimens' fragmentary nature. When more complete material is recovered, some significant contrasts with the Hadar/Laetoli sample should emerge and taxonomic revision may be required, as with the Omo specimens below.

5.3.5   *Omo*

Recognition of *A. afarensis* and the recovery of the 'black skull' have led to reappraisal of the fragmentary hominid fossils from the lower Omo basin, south Ethiopia, where extensive exposures provide a stratigraphic record for much of the Pliocene and Pleistocene. They were investigated in the late 1960s by American, French and Kenyan expeditions led by Howell,

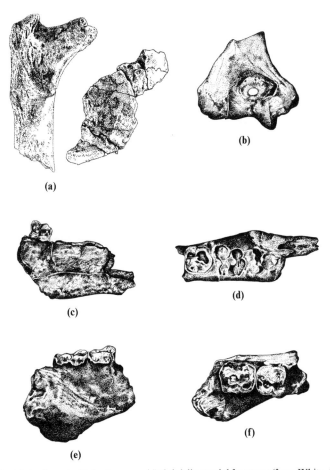

Figure 5.8 (a) *A. afarensis*: Maka femur and Belohdelie cranial fragment (from White, 1983). Early East African hominids. (b) Kanapoi humerus (c, d) Lothagam mandible. (e, f) Tabarin mandible (from Hill and Ward, 1988).

Coppens and Leakey, recovering much environmental and faunal evidence as well as hominid fossils and archaeological sites.

Shortly afterwards Richard Leakey and associates explored sites on the eastern side of Lake Turkana (then Lake Rudolf) which overlap in part with those of the Omo succession (see chapter 6). A major triumph of recent years has been the integration of these separate sequences through tuff correlations, to provide a stratigraphic framework for much of the region, with the Omo sites linking the Awash with those of northern Kenya and making for more precise dating over a wide area (Feibel *et al.*, 1989). The Turkana sites are considered in more detail in chapter 6.

The Omo sequence is divided into six formations, of which three—the Usno, Shungura and Kibish formations—contain hominids. The Usno exposures, which are limited in extent, are early (*c.* 4–3 mya), while the Kibish deposits are of recent (< 0.2 mya) date. The bulk of the sequence consists of extensive (> 60 km) exposures of fluvial, delta and lacustrine sediments of the Shungura formation, averaging 700 m, and in places > 1000 m thick. The formation consists of 12 members (A–L), each delimited by a tuff at its base and with further ash layers within, which form the basis for a potassium–argon chronology extending from *c.* 3.5 to *c.* 1.0 mya.

Hominid remains, fragmented by river action, are known throughout the sequence. The earliest are teeth from the White Sands and Brown Sands localities of the Usno formation, correlating with Shungura B (i.e. just over 3 my old). They consist, with one exception, of cheek teeth and were originally assigned to *A. aff. africanus*, although the unusually small size of

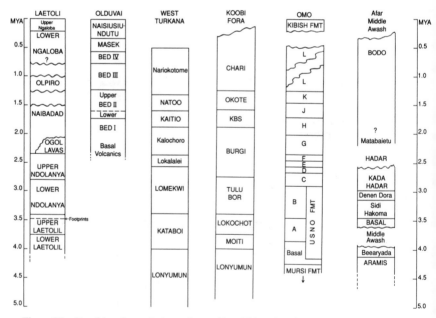

Figure 5.9   Provisional correlations of some East African hominid site sequences (based on Brown and Feibel, 1986; Leakey and Harris, 1987; Feibal *et al.*, 1989).

the crowns was noted. They are now considered, on both morphology and proportions, to represent *A. afarensis*, as do other, slightly younger (2.9 my old) teeth from Upper Shungura B.

5.3.6 A. aethiopicus

There are further isolated teeth from member C (2.8–2.5 mya) together with a partial, strongly built mandible lacking tooth crowns but, to judge from the roots, with broad cheek teeth and sizeable canines. On its discovery a new taxon, *Paraustralopithecus aethiopicus*, was created for this specimen (Arambourg and Coppens, 1967) and then abandoned, with the material referred to *A. aff. africanus*. With the evidence of the 'black skull', the member C fossils are now considered early members of the East African robust clade (Howell, 1987). If these are to be specifically distinguished from later *A. boisei* (see above), then the species *A. aethiopicus*, based on the Shungura C mandible, has priority. Further remains of this lineage may be among the fragmentary remains (teeth, a partial humerus) from member D and, more certainly, those from members E and F (2.4–2.3 mya) which include postcrania (ulna, calcaneum), parts of a juvenile cranium and two partial mandibles. The first evidence of typical *A. boisei* with small anterior and massive cheek teeth is provided by two mandibles and isolated teeth from lower member G (2.2 mya).

Also from Omo members F–G are isolated teeth and a fragmentary cranium referred to *Homo* sp.; these are discussed further in chapter 6 (p. 121), as are archaeological sites within the Shungura formation (p. 137).

5.4 Cranio–facial biomechanics

In a remarkable study Rak (1983) analysed australopithecine facial morphology in great detail, adding appreciably to our understanding of the biomechanical significance of many features, pointing up interspecific differences, and providing evidence from which to infer ecology and dietary adaptations. Rak considers early hominid facial morphology to have been primarily influenced by several factors: retraction of the palate under the face rather than projecting in front; an anterior shift of the zygoma and the origin of the masseter; and expansion of the anterior portion of the temporalis muscle. These interact with each other and with the expansion of the cheek teeth, especially the molarisation of the premolars, seen in early hominid species.

Rak views *A. afarensis* as possessing the most primitive hominid facial morphology, sharing many similarities with the African apes. The palate is long, flat and shallow, the anterior teeth comparatively large, and the cheek teeth moderately sized. The premaxillary region (the naso-alveolar clivus) is strongly curved and forms a projecting snout that is well separated from the lateral face by a distinct canine fossa. The margins of the nasal aperture are thin and sharp, as in modern hominoids, with no evidence of reinforcement. The cheek region below the orbit is braced by a transverse bony buttress, below which the bone slopes downwards and backwards, parallel to the action of the anterior masseter muscle. The lower border of the maxillary zygomatic process is thin and rises relatively far back, so bracing the alveolar bone in the molar region. The supraorbital torus joins smoothly with the frontal and projects laterally as a wide triangular bar.

The *A. africanus* face, while still prognathous, is not so projecting and the palate lies further underneath the face. The cheek teeth, especially the premolars, are larger, and strong occlusal forces extend further forward along the tooth row of the relatively deep palate. To accommodate these the anterior face is reinforced by the anterior pillars bounding the nasal aperture and converging in the interorbital area. The premaxillary region is flat, not curved, and together with the pillars forms the naso-alveolar triangle projecting forward from the lateral face. The infraorbital plate also extends forward, and the strongly built zygomatic process is inclined more or less vertically so that the transverse buttress is obscured. So too is the canine fossa, the maxillary furrow behind the pillars being all that remains. Laterally the cheek swells to a distinct boss, the zygomatic prominence, so extending the masseter anteriorly and laterally before swinging sharply backward to form the zygomatic arch. The anterior portion of *temporalis* is also well developed, and the supraorbital torus laterally forms a narrow bony bar.

Many of these features are further developed in *A. robustus*, while other aspects of its morphology can be viewed as adaptations to resist the greater forces generated by the enlarged cheek teeth. The palate is further retracted, producing a flatter facial profile, and the cheek region advanced anteriorly and extended laterally to give a diamond shape to the face in frontal view. The anterior pillars are especially well developed and more vertically orientated than in *A. africanus*, and because of palatal retraction the premaxillary region is set within these, so forming the naso-alveolar gutter rather than projecting anteriorly.

The advancement of the infraorbital bone of the cheek, which slopes forwards and downwards, obscures the maxillary furrow except near its

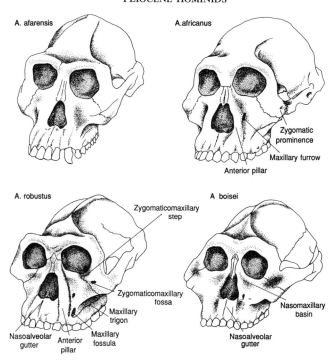

A. afarensis

A.africanus

Zygomatic
prominence

Maxillary furrow

Anterior pillar

A. robustus

Zygomaticomaxillary
step

A boisei

Zygomaticomaxillary
fossa

Maxillary
trigon

Nasomaxillary
basin

Maxillary
fossula

Nasoalveolar   Anterior
gutter   pillar

Nasoalveolar
gutter

Figure 5.10 Australopithecine skull anatomy as interpreted by Rak (1983). For further
details see text.

base close to the alveolar border. Instead there is a shallow depression
before the bone swells and thickens along the line of the suture between
maxilla and zygoma to form the zygomaticomaxillary step, reinforcing the
cheek against the torsional effects of the powerful masseter as it contracts.
This is set further forward and more laterally than in *A. africanus* because of
the characteristic facial dishing of *A. robustus*, with the mid-face in a central
hollow and the most lateral parts beyond the steps projecting anteriorly.
The zygomatic root is so far forward (over P^3/P^4) that the mid-
rear tooth row is no longer braced, and there is massive thickening of the
anterior part of the palate, analogous to the symphysial tori of the
mandible, to withstand the shear stresses generated in the unsupported rear
portion of the palate during chewing. Postorbital constriction is marked and
the temporal lines and crest show the temporal muscles to be large and
situated well forward on the rear of the supraorbital torus. This acts as a
cross-beam, strongly built up medially in the glabellar region (where it is

subject to compressive forces during chewing) and tapering laterally to an isolated bar (because of the postorbital constriction) and under tension as a result of the masticatory muscles' contraction.

A. *boisei* displays further retraction of the palate and a diamond-like facial shape from the front, reflecting its great breadth at the level of the cheek bones and the orientation of the maxillary zygomatic process (sloping upwards) and the supraorbital torus (sloping downwards) from the pronounced glabellar area. The premaxilla is similar to A. *robustus* and the nasal aperture has blunt margins, but no anterior pillars or zygomaticomaxillary steps are evident. Instead the bone below the orbit slopes downwards and forwards on the same plane as that around the nasal aperture to form a plate which extends laterally before curving back into the strongly flaring zygomatic arch; the general effect is that of a curved visor. The zygomatic root is more anteriorly situated than in other hominids (over $\underline{C}/P^3$), and its narrow margin is remarkably thin, adding to the impression of a plate. Rak suggests that the infraorbital region has extended so far laterally and anteriorly that it can no longer be braced by the crossbar of the supraorbital torus, and the curved visor provides a free-standing rigid structure. The need for this is reinforced by the depth of the infratemporal fossa to accommodate the very large temporal muscle; extreme postorbital constriction means that the lateral part of the supraorbital torus is a thin bar unsupported by the braincase wall. The extreme lateral development of the zygoma also shifts the masseter laterally, so that its contraction threatens to pull the visor and arch downwards and outwards, tearing apart the thin supraorbital torus. These forces are countered by heightening the glabellar keystone with the remainder of the torus sloping inferolaterally, instead of being a transverse horizontal bar, while at the other end of the arch support is provided by the extensive overlap of the temporal and pariental bones along their suture.

Rak considers A. *afarensis* and the two South African species as structural variants of the general early hominid face, and A. *boisei* with its visor as representing a novel form developed from the others. He identifies a morphological trend, A. *afarensis* → A. *africanus* → A. *robustus* → A. *boisei*, and also views this as a phyletic sequence, with A. *africanus* representing an early stage in the development of the robust morphology. Moreover, since Rak points to similarities between the faces of A. *afarensis* and early *Homo*, his analysis provided major support for the theory of Johansen and White of A. *afarensis* as the stem hominid directly ancestral to *Homo*. However, the 'black skull', with its early appearance of the A. *boisei* visor, casts doubt on

Rak's interpretation of this morphology as being a relatively late development derived from that resembling *A. robustus*. It is perhaps better to view the *A. robustus and A. boisei* facial morphologies as alternative structures for withstanding powerful masticatory forces rather than one developing from the other. Other aspects of this phyletic (as opposed to structural) interpretation are discussed below, but Rak's biomechanical analysis provides valuable insights into the factors influencing australopithecine cranial morphology.

5.5 Postcranial material

Australopithecine pelvis and lower limb remains reveal a clear pattern of bipedality; overall they are more human than ape-like, while possessing distinctive features that make for a unique morphology not matched in any modern primate. The South African and Hadar pelves are small and noticeably shorter than pongids, with wide, fan-shaped ilia that curve less than modern humans and so flare laterally. The iliac pillar is not distinct, and instead the anterior portion of the iliac blade is thickened; the anterior inferior iliac spine is prominent, so giving advantage to rectus femoris, an important flexor of the thigh and, in the South African specimens at least, a well-developed ilio-femoral ligament is present, preventing hyperextension of the thigh on the hip. The pubis is relatively long and the ischium short, with the hamstring moment arm, as revealed by the position of the ischial tuberosity, similar to humans, so aiding thigh extension. The acetabulum (hip socket) is small, as is the sacral surface, and the distance between these is greater than in humans, suggesting rather less efficient transmission of trunk weight to lower limb.

The femoral head is small (so matching the evidence of the hip socket) and the neck long, while the trochanters are prominent but do not extend beyond the line of the shaft as they do in man. At the distal (knee) end the shaft is set obliquely to the articular surfaces in a valgus ('knock-kneed') position, so bringing the knee directly under the hip joint. The lateral femoral condyle is larger than the medial one and the intercondylar notch and patellar groove deep, as in modern humans. Overall the femoral morphology indicates a locking knee joint and effective weight transmission down the limb.

On the basis of this pattern most workers accept that the australopithecines were bipedal, although perhaps not exclusively so, with other modes as part of the normal locomotor range. Early accounts were largely based upon the incomplete (and partly distorted) South African material, and emphasised contrasts between *A. africanus* and *A. robustus*, with the latter apparently less well adapted to bipedalism. More extensive and undistorted remains from East and South Africa point to a common set of features in australopithecine postcrania (with the possible exception of *A. afarensis*, see below), with no evidence of major interspecific contrasts, all species being well adapted to bipedalism.

The exact nature of that bipedalism is less clearly defined. Some workers regarded australopithecine locomotion as indistinguishable from our own, while others (e.g. Zihlman, 1989) view it as less energetically efficient, and perhaps associated with other positional behaviours (Rose, 1984b). Some, e.g. Leutneggar (1987), have suggested that contrasts in pelvic and femoral morphology between *Australopithecus* and modern *Homo* reflect contrasts in neonate brain size rather than locomotor differences (see p. 97–98). Lovejoy (1974; 1988) has argued that the small brain of *Australopithecus* allowed it to be more, not less, efficient bipedally than modern humans, in whom head size has affected pelvic proportions, especially internal diameter and the distance between the hip joints. Lovejoy regards the more laterally flared ilia and longer femoral neck of *Australopithecus* as giving greater mechanical advantage to the abductors (gluteus medius and minimus) in stabilising the trunk over the weight-bearing leg while the other leg is swinging forward during walking.

The AL 288 skeleton provides information about postcranial proportions not otherwise available. Lucy's hindlimb was short both absolutely and relative to forelimb and trunk lengths; as such it affected stride length and walking velocity. Moreover, Jungers (1988a) has shown that Lucy's hindlimb joint sizes, although rather bigger than those of apes, are much smaller than modern humans even when differences in overall body size are taken into account, implying less complete adaptation to full weight bearing by the hindlimb.

The anatomical evidence suggests australopithecine ground locomotion was certainly bipedal, but not necessarily of modern human form.

Analysis of the postcrania inevitably involves inferring locomotor behaviour from morphology, but more direct evidence exists in the form of the 3.6 my-old Laetoli footprints. There are three tracks recognised, differing in size but with prints essentially modern in outline, with adducted big toe. The prints reveal the arched structure of the early hominid foot with well-developed medial and lateral longitudinal arches, tripod arrangement for weight distribution (big toe, little toe, heel) and a toe-off/heel-strike activity pattern. Locomotor reconstructions based on the tracks suggest a

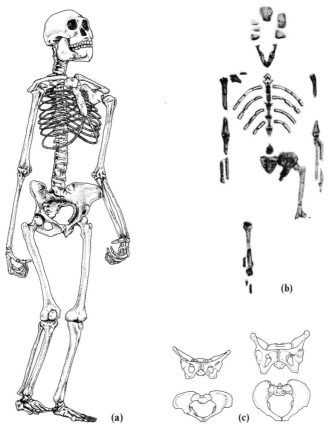

Figure 5.11 Early hominid locomotor system. (a) *A. afarensis* ('Lucy') walking as re-constructed by Lovejoy. (b) AL 288 ('Lucy') partial skeleton. (c) Front and upper views of AL 288 pelvis (left) compared with modern human. Note flared, uncurved iliac blades and shallow dorso ventral diameter of AL 288 pelvis (after Tague and Lovejoy, 1986).

relatively short, but not necessarily slow, stride pattern to early hominid bipedalism.

The overall picture is therefore clear enough, but difficulties remain, especially with the *A. afarensis* material. The Laetoli prints are short-toed, while the Hadar foot remains have long, curved phalanges, leading Tuttle to doubt whether they could have made the prints. White and Suwa (1987) scaled the larger Hadar foot remains to Lucy's size and demonstrated a close fit with some of the Laetoli prints.

Stern and Sussman (1983) have drawn attention to a set of anatomical features in the shoulder and trunk skeleton, hip bone, femur, lower leg, ankle and feet of some of the Hadar specimens, especially Lucy. These they interpret as indicating wider ranging limb movements to permit adept arboreal activity, and less than modern bipedalism. The larger Hadar postcrania do not show these features to the same extent, and so Stern and Sussman conclude that there were significant behavioural differences between the sexes in *A. afarensis*, with females appreciably more arboreal than the larger bodied males. Since the morphological contrasts cited in support of this interpretation are not found in even the largest sexually dimorphic modern pongids, Stern and Sussman conclude that 'the degree of sexual differences in locomotor behaviour in *A. afarensis* was greater than in any living apes'. However, at least some of the features noted by these authors, e.g. femoral head form (Asfaw, 1985), have been discounted by recognising that they exhibit high variability within modern *H. sapiens*, without necessarily being related to locomotor differences.

Senut and Tardieu (1985) also argue for a division of the Hadar postcrania on the basis of their studies of the elbow and knee joints. They recognise two patterns: one of lax/flexible elbow and stable knee joints (larger specimens) attributed to *Homo*, and one of stable elbow and lax knee joints (in some smaller specimens including Lucy) that they assign to *Australopithecus* and consider indicative of climbing ability. McHenry (1986) argues that the contrasts in knee joint morphology revealed by distal femoral morphology have been overdrawn, and are in part size dependent. Thus postcranial contrasts are interpreted by Senut and Tardieu (1985) as indicating specific (and even generic) distinctiveness, whereas Stern and Sussman (1983) accept the integrity of *A. afarensis* and regard them as resulting from sexual dimorphism, although their conclusion has been described by Day (1985) as 'special pleading for a wider range of Hadar sexual dimorphism than previously known in primates'.

While some of the arboreal indicators of the Hadar postcrania are paralleled elsewhere, e.g. the Sterkfontein scapula fragment and humerus (Vrba, 1979) had a cranially directed shoulder joint and powerful upper arm suitable for suspensory activity, there are no specimens from other sites with the set of features suggesting a distinct mode of bipedality as with Hadar *A. afarensis*. Indeed, Stern and Sussman (1983) stress that in their view the South African hominids were more committed terrestrial bipeds. McHenry (1986) emphasises the fundamental similarity of *A. afarensis* and *A. africanus* postcrania, and considers that the differences, although much debated, are trivial. 'Except in relatively major details, the postcranium of

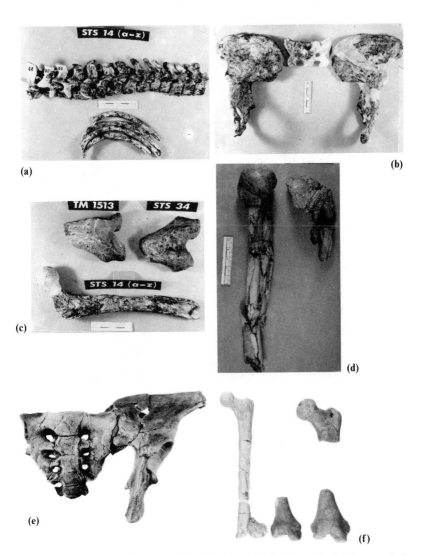

Figure 5.12 *Australopithecus* postcrania. *A. africanus*: (a) Sts 14 vertebral column and rib fragments. (b) Sts 14 pelvis. Note prominent anterior iliac spines and lack of curvature of ilium. (c) Sts 14 femoral shaft compared with distal (knee) femoral fragments. Note size contrasts. (d) Sts 7 scapula fragment and humerus. Note size and muscle impressions on humerus. *A. afarensis*: (e) AL 288 sacrum and left innominate. Note poorly marked anterior inferior iliac spine and lack of curvature of ilium. (f) Hadar femora, left to right: AL 288, AL 129A, AL 333- 3 and 4. Note size contrasts and lack of 'flare' at upper end of femoral shaft reflecting unexpanded trochanters.

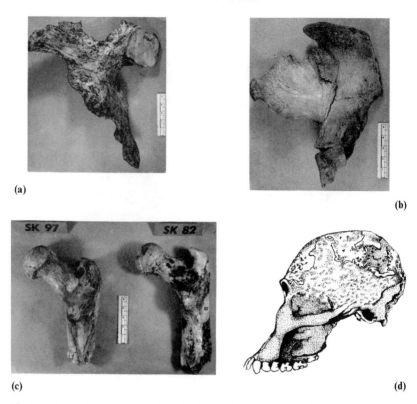

Figure 5.13 *Australopithecus postcrania. A. robustus* (a) SK 80 hip bone (incomplete and distorted). (b) SK 3155 hip bone. Note great breadth of ilium. (c) Swartkrans femora. Note small femoral heads, long necks and straight-sided femoral shafts reflecting unexpanded trochanters. (d) *A. africanus* cranium Stw 252. This specimen significantly extends cranio-dental variation in the Sterkfontein sample. Note the degree of alveolar prognathism and the large size of the incisors and canines (after Clarke, 1988).

the first bipeds, *A. afarensis* and *A. africanus* are very similar to one another ... and ... implies that locomotor and postcranial adaptations were very similar and unlike any living form'.

None of the available postcranial fossils provides any support for the evolution of hominid bipedalism from a knuckle-walking ancestor; this, together with the pattern of size variation of the fossils, interpreted as sexual dimorphism, militates against a chimpanzee-like form, especially the pygmy chimpanzee, *P. paniscus*, as a likely ancestral morphotype.

5.6 Growth, development and body size

Because of their hominid status, initial investigation of australopithecine growth and development tended to focus on human models as analytical aids. For example, in an influential study of australopithecine demography, Mann (1975) considered the South African species to have followed a modern human pattern of development, with a prolonged growth period. The larger samples and new techniques of study now available have radically altered this view.

Consistent with their small brains, head size in newborn australopithecenes appears to have been no bigger than that of chimpanzee neonates or even less. Lucy's pelvis, although broad, is very shallow dorsoventrally, making for obstetric difficulties even if neonate head size were only that of the chimpanzee (Tague and Lovejoy, 1986), so suggesting a size even smaller (Leutnegger, 1987). The Sterkfontein Sts 14 pelvis is bigger and could accommodate a chimp-sized neonate (Leutnegger, 1972), but none of the available evidence points to a 'helpless' newborn of the modern human kind.

Data on australopithecine infancy and childhood are consistent with this picture. Enamel increments (pp. 32–34) provide a more precise method of calibrating tooth development than was available to Mann, and have been studied by Beynon and Dean (1988). Their results indicate periods of dental growth appreciably shorter than in modern humans, and estimated ages at death of immatures much lower than Mann's figures. All australopithecines studied appear to have completed tooth development within similar periods, which are relatively short and similar to apes—there are no indications of an extended growth period.

Beynon and Dean's study reveals significant differences of pattern between *A. afarensis/africanus* and *A. robustus/boisei*. In the former group the large lower incisors develop slowly, and are still not complete when M_1 occludes (*c.* 3.5 years); in the robusts incisor formation and occlusion are rapid, occurring at the same time as M_1 occlusion, which at 3 + years is slightly earlier than in *A. afarensis/africanus*. These figures compare with ages of 3–4 years in great apes and 6–8 years in humans for M_1 occlusion, which has been shown (Smith, 1989) to correlate very highly with brain weight, which has reached 95% of adult value at that stage. The early estimates for M_1 eruption are therefore consistent with the small adult brain sizes of *Australopithecus*.

Current studies thus suggest a *period* of australopithecene growth and development similar to apes, not humans, but with significant differences of *pattern* between early hominids and modern species. Later developmental events, e.g. sexual maturity and adult age at death are frankly speculative, but if the sequence and timing of second and third molar eruption were similar to apes, there are clues pointing to a short lifespan. The South African fossils display marked dental wear, with tooth crowns completely worn away in older individuals. Significant wear is already on M^1 and other teeth before M^3 are fully occluded (8–12 years in apes), pointing to an estimate of late teens/early twenties for age at death when starvation and enfeeblement would have followed tooth destruction. This in turn suggests an early (?8–? years) onset of sexual maturity if population numbers were not to dwindle rapidly to the point of extinction within a very few generations.

Inferred differences in body size between australopithecine species are a complicating factor when comparing gracile and 'robust' forms. Size differences inevitably produce shape contrasts through scaling effects, and some workers, e.g. Pilbeam and Gould (1974) therefore used allometric models in attempts to 'subtract' those differences in proportion that follow from changes in overall size from contrasts which reflect size-independent structural changes. Most studies have concentrated upon dentition, where sample sizes are largest; detailed conclusions differ, but most indicate that, while the enlarged cheek teeth of *A. robustus/boisei* may reflect greater body size, this is unlikely to account for the relatively small anterior dentition of robusts. In addition, cranial studies (Corruccini and Ciochan, 1979; Rak, 1983, Bilsborough and Wood, 1988) suggest that certain aspects of facial morphology in *A. robustus/boisei* are not a simple consequence of greater size.

In fact, the larger fossil samples now available indicate marked *intra*specific size variation and considerable *inter*specific overlap, with between-species differences much less than previously supposed. Two recent independent studies (Jungers, 1988b; McHenry, 1988) relating lower limb dimensions (joint surface dimension, femoral shaft width) to body size in a variety of primates have been used to assess australopithecene body weights. Jungers gives two estimates: one based on all hominoids, including humans, who have unusually large lower limb joint surfaces for a given body size, the other excluding humans, which probably gives a closer estimate (Table 5.2).

McHenry's study, based on femoral shaft diameter, includes a human sample and derives body weights very similar to Junger's all-hominoid values. The australopithecine body size ratios (smallest to largest) can all be accommodated within those of the great apes, indicating that claims of

Table 5.2 Estimated body weights (kg) of *Australopithecus* species (Jungers, 1988b).

	All hominoids		Excluding humans	
	Range	Mid-point	Range	Mid-point
A. afarensis	30–68	49	30–81	56
A. africanus	33–58	45	33–68	50
A. robustus	37–58	47	42–89	65
A. boisei	33–69	51	37–89	63

species heterogeneity, e.g. at Hadar, based on *size* are unsupported, although body size diversity would certainly be greater than that represented by the very small samples and there still remain contrasts in morphology to be explained (see above).

Pronounced sexual dimorphism is the most likely explanation for the intraspecific ranges, and has implications for reconstruction of australopithecine social organisation and behaviour (see below). These studies also point to other conclusions: they confirm that *all* australopithecine species have large cheek teeth compared with general primate models, but also indicate, contrary to earlier analyses, that their further enlargement in *A. robustus/boisei cannot* be explained by increased body size in these forms. In the same way, the rather larger cranial capacity of the robusts seems to represent a relative increase in brain size, and is not a simple consequence of body size differences. On these data 'gracile' and 'robust' are potentially misleading terms that at best 'refer to craniodental features only' (McHenry, 1988).

5.7 Evolutionary relationships

The recognition of australopithecine species—initially *A. africanus* and *A. robustus*, with *A. boisei* as a later addition—led to much debate as to which form (irrespective of dating) was the more primitive. Robinson (1968; 1972) and Jolly (1970) argued that *A. robustus* retained more primitive features, but most workers agreed with Tobias (1967) in viewing *A. africanus* as the more primitive form from which the *A. robustus/boisei* morphologies were derived. Various early cladistic analyses (e.g. Delson *et al.*, 1977; Olson, 1978) considered *A. africanus* primitive and *A. robustus* to be derived, while Wallace (1975; 1978) considered *A. africanus* primitive on non-cladistic grounds.

Over the last decade these issues have been overshadowed by the recognition of *A. afarensis*, and the discovery of KNM WT 17000. Johanson and White (1979) and White *et al.* (1981) viewed *A. afarensis* as a morphologically primitive stem hominid, ancestral to both other australopithecine species and to the genus *Homo*, with *A. africanus*, until then considered primitive, as an early stage in the differentiation of the *A. robustus/boisei* lineage, and so excluded from human ancestry.

The thrust of White *et al.*'s argument was to demonstrate the more primitive nature of *A. afarensis*, and the greater similarity of *A. africanus* to *A. robustus/boisei*. In their view it was the presence of derived 'robust' features in *A. africanus* that excluded it from human ancestry, and the absence of these in *A. afarensis* that identified the latter as the common ancestor, rather than any very close resemblance between *A. afarensis* and *Homo*. The only claims for similarity between *A. afarensis* and early *Homo* were in a limited set of dental characters (especially relative proportions of anterior and cheek teeth) and associated masticatory features. Their interpretation received major support from Rak's study (1983) of early hominid facial anatomy in which he identified a morphological and phyletic trend from *A. afarensis* → *A. africanus* → *A. robustus* → *A. boisei*, and the Johanson and White scheme was accepted by many workers.

Nonetheless, other phyletic interpretations were proposed. Tobias (1981; 1987) argued that *A. afarensis* is merely an East African subspecies of *A. africanus*, while Wood and Chamberlain (1986) suggested *A. afarensis* was linked with the robusts, and *A. africanus* with *Homo*. Skelton *et al.* (1986) and Bilsborough (1986) argued (on different grounds) that *A. afarensis* and *A. africanus* were chronospecies of a single clade, which split between robusts and *Homo* after *A. africanus*.

Prior to the recovery of the 'black skull' there were thus two main phyletic schemes: Johanson and White's, with *A. afarensis* as the last common ancestor of other australopithecene species and *Homo*, and with *A. africanus* as a proto-robust form; and Tobias's with *A. afarensis* as an East African subspecies of *A. africanus*, which is the last common ancestor of *Homo* and robusts. Skelton *et al.* (1986) refined Tobias's scheme by noting that certain features indicated the split between *Homo* and *robustus* to have been after the known *A. africanus* specimens.

The recovery of the Lomekwi KNM WT 17000 skull, broadly resembling *A. boisei* in overall morphology but with *A. afarensis*-like anterior dental proportions, throws both these schemes into disarray and introduces a very large spanner into the phyletic works. Its indication of the early (2.5 + mya) appearance of a robust clade has important implications: it undermines Rak's interpretation of a phyletic trend in facial morphology (see above)—an important buttress to the view of *A. afarensis* as the most primitive, generalised species. More particularly, it undermines Johanson and White's argument for a close link between *A. afarensis* and *Homo* on the basis of dental proportions, for these are now seen in an otherwise *A. boisei*-like skull, suggesting that they either evolved in parallel in this form or were widely distributed across several early hominid species (i.e. were 'primitive'). On either count there is no reason to view them as indicating especially close links between *A. afarensis* and early *Homo*.

While *A. afarensis* is still regarded by many as a primitive, stem hominid, it is cast in that role more by virtue of being 'in the right place at the right time'. Its relationship to *A. africanus* is probably that of antecedent chronospecies rather than separate clade (Bilsborough, 1986; Boaz, 1988) and re-establishes the latter species as likely immediate ancestor of the genus *Homo*. On the other hand KNM WT 17 000 would appear to exclude *A. africanus* from the ancestry of *A. boisei*, if not *A. robustus*, and has other important phyletic implications.

The features of the 'black skull' suggest derivation from *A. afarensis* or something like it, and so *A. africanus* cannot be the sole ancestor of all robusts, a view accepted by Kimbel *et al.* (1988b). On the other hand, there is still a case for deriving South African robusts from *A. africanus*, some specimens of which are distinctly *robustus*-like in their features. If this is so, it means that the robusts do not form a single clade, but are parallel developments. Wood (1988) has pointed out that many similarities between *A. robustus* and *A. boisei* relate to the masticatory system, and thus may especially arise through convergence. Moreover, the developmental sequence of some apparently similar features, e.g. multirooted premolars, suggests that they may be analogous traits, not homologues.

However, there are many other common features, e.g. flexed basicranium, structure of the ear region, brain enlargement and proportions, cerebral circulation, tooth development and thick enamel, that are difficult to accept as parallelisms, and point to monophyly for the robust group. This is accepted by many workers who continue to consider, on balance, that robusts form a single clade (Grine, 1988a; Wood, 1988). Many of the above features are either not present or not fully developed in WT 17 000, suggesting that it represents an early phase in robust evolution with a period of monophyly between South and East African robusts after 17 000, and before the split into geographically distinct species. On the other hand, if this is correct, both morphology and dating make it difficult to locate *A. africanus* on the direct ancestry of even the South African robusts, and the

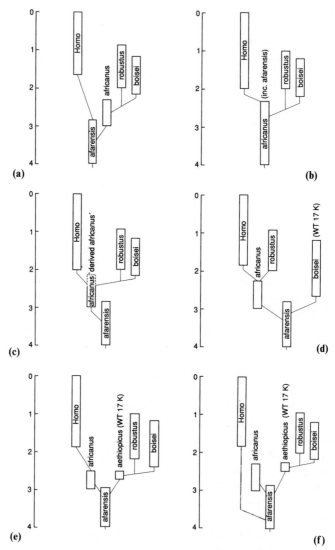

Figure 5.14 Some early hominid phylogenies. (a) Johanson and White, 1979; (b) Tobias, 1981; (c) Skelton *et al.*, 1986; (d) Walker and Leakey, 1988; (e) Kimbel *et al.*, 1988a; (f) Kimbel *et al.*, 1988b (format modified from Grine, 1988).

claims for a *phyletic* sequence Sterkfontein → Kromdraai → Swartkrans must be seriously awry.

Some of the similarities (synapomorphies) between the two robusts such as brain enlargement and form, cerebral circulation and basicranial flexion are also shared with *Homo*. Unless these are also parallelisms, they imply a period of common ancestry for *A. boisei/robustus* and early *Homo* that is after *A. afarensis/A. africanus* since these do not possess such features. The period is unlikely to have been a long one, given the many contrasts between robusts and early *Homo* in face, braincase, tooth proportions, enamel thickness, etc.

In fact it is not possible to construct a phyletic model for this tangle of species that is compatible with all current evidence without invoking convergent/parallel evolution and/or reversals in at least some characters. In other words, some similarities must be homoplasies; the problem is to determine which ones. In such situations the usual solution is to invoke the principle of parsimony, 'Occam's razor', to select the phyletic scheme with fewest parallelisms. Even this creates difficulties, for the parallelisms identified vary from study to study and may depend, for example, upon whether size-corrected or unadjusted data are used; the number and nature of the outgroups; character selection, particularly whether dental traits predominate over non-dental in the study; and whether separately identified characters are, in fact, highly correlated. There is growing evidence over many groups that parallel evolution is a frequent occurrence, and early hominids are unlikely to have been an exception. As with so much else in palaeoanthropology, more evidence is needed. My own preference is for Figure 5.14 (e), followed by (d), while recognising the difficulties referred to above.

The above schemes, despite all their uncertainties, have in common their recognition of several specific morphologies; whatever detailed adjustments may need to be made there is now general recognition that early hominid evolution was certainly polyphyletic. The possible adaptive basis of such diversity is discussed below.

5.8 Australopithecine environments and adaptation

A major cold phase in the deep sea core record *c.* 2.5–2.6 mya has been associated with cooler, more arid conditions over much of Africa (Prentice and Denton, 1988) causing major ecological shifts and increased faunal turnover (Vrba, 1988). Some see the appearance of robusts in South and

East Africa as one aspect of such an evolutionary pulse, and their extinction has been linked (Klein, 1988; Vrba, 1988) to a later cold phase 0.9 mya (see below and chapter 6). However, WT 17 000 shows that many 'robust' features had already evolved by *c.* 2.5 mya, and White (1988) sounds a timely note of caution: as more climatic events are recognised, so the probability increases that some will coincide with hominid evolutionary events, without implying causation.

On a more local scale, geological, faunal and other data yield information on early hominid environments, which in turn may serve as the framework for behavioural reconstructions. However, there is growing awareness that such information is highly tentative. Knowledge of early hominid sites is severely skewed—limestone cavern in-fills in South Africa, major river systems or lakeside localities within Rift basins in East Africa, with few other contexts reported. Taphonomic and depositional agencies (e.g. river channels that rework and mingle specimens originally widely separated in space and time) and fieldworker bias are further distorting agencies, so that the available evidence is most decidedly *not* an unbiased sample of past environments (White, 1988). With these caveats in mind, it is worth summarising the contextual associations of the australopithecine radiation.

Geomorphological and faunal data from South Africa indicate that the *A. africanus* sites accumulated during wetter, more wooded conditions than those associated with *A. robustus*. The member 3 fauna at Makapansgat suggests mixed bush and woodland, and similar conditions probably obtained at other *A. africanus* sites where the climate was wetter than now, and significantly more so than during the time of Swartkrans member 1. Butzer (1974) considered the Taung deposit to have been laid down in subhumid/humid conditions, whilst analysis of bovid faunas (Vrba, 1974) indicates greater tree cover during the time of Sterkfontein member 4 than Swartkrans member 1, which appears to have been predominantly open grassland, although there were some wooded areas near the locality. The Kromdraai fauna suggest an environment intermediate between the rather wetter *A. africanus* sites and the drier Swartkrans member 1.

In East Africa the pattern is broadly similar, although Laetoli is an exception. Here the environment appears to have been relatively dry and open, with the site itself close to water, perhaps a pond or water hole. At Hadar data indicate bush/woodland and marshy lake edge environments in the lower part of the succession (basal and lower SH members), giving way to more open conditions. In upper SH/lower DD members lake expansion resulted in flood plain conditions with grassy expanses but with bush and wooded areas along streams. The fauna suggests continuing mixed habitats in upper DD times with expansion of open grassland, a process continuing into the lower KH member (Johanson *et al.*, 1982a).

The Omo deposits, the lower horizons of which partly overlap with Hadar, have yielded *Australopithecus* with evidence of closed and/or open woodland and shrub cover. Higher levels (member E and above) provide some evidence of drier, more open conditions with expanded grassland and some shrub cover but little woodland, a shift associated with the appearance of robust australopithecines and *Homo* in the sequence. Throughout the sequence the river was a large one, bordered by forest, which probably acted as a refuge area, buffering species against the environmental changes away from the river (Vrba, 1988).

At Koobi Fora and Olduvai the evidence again indicates robust australopithecines and *Homo* in open, semi-arid environments, although with limited tree cover near water sources. Environments at these sites are discussed further in chapter 6, pp. 110–111, 114–117, 135–137. A complicating factor in East Africa is tectonic activity which affected drainage patterns, the effects of which could be marked without necessarily implying climatic change. Overall the evidence suggests that all australopithecines inhabited broadly open or mixed environments but with later, robust forms in drier habitats than the earlier *A. africanus* and *A. afarensis*, although with tree cover near water sources where most fossils are located.

A study of robust sites by Shipman and Harris (1988) gave results that conflict with this picture. They used frequencies of selected bovids as habitat indicators, concluding that '*A. boisei* probably preferred closed over open habitats and wetter rather than drier habitats'. Their approach has been severely criticised by Vrba (1988) for not including all bovids, so distorting results, and, for example, precluding discrimination between South African sites, all of which are categorised as 'open arid'. Closed habitats for *A. boisei* would be compatible with tooth wear indicators (see below) and suggest ecological/adaptive contrasts between South and East African robusts. However, it is difficult to reconcile Shipman and Harris's results for some sites with the evidence of other habitat indicators at those sites, and widespread closed wet habitats are probably unrealistic. The later East African sites are further discussed in chapter 6.

Early behavioural reconstructions of *Australopithecus*, based on the South African evidence, saw them as 'proto-humans' with home bases, and a hunter–gatherer subsistence pattern based on technology. Whilst understandable in light of the need to establish their hominid status, such reconstructions are now known to be well wide of the mark. Australo-pithecines were not human in any accepted sense of the term, and their ecology and behaviour are unlikely to be closely matched by any living primate.

This shift of attitude can be dated to Brain's demonstration (1970) that the Swartkrans faunal assemblage was the result of carnivore predation, not hominid activity. This conclusion has been extended to the other South African sites, and it is now generally accepted that there is no evidence that the caverns were living sites, or that *Australopithecus* was a systematic predator. Carnivore predation (especially by sabre tooths and leopards) is also consistent with the parts of the carcass represented (skulls and other non-meaty parts) and the fact that most remains are of young and immature (and therefore weak and inexperienced) or old and decrepit individuals, enfeebled by starvation following the destruction, through wear, of their dentition. There appear to be very few young mature males represented in the South African collections.

Following rejection of the notion of *Australopithecus* as a 'killer-ape', increased attention was directed towards the importance of plant foods for early hominids, although there are obvious difficulties. The flexibility of modern primates suggests that models such as the 'seed-eating' hypothesis (Jolly, 1970) that stress a particular food item are likely to be inadequate, and there are few plant remains preserved at African Plio-Pleistocene sites.

Peters and O'Brien (1981) attempted to define the *potential* plant food niche of early hominids by considering those foods known to be utilised by chimps, baboons and modern hunter–gatherers. There is great variety, but predominant items are fruit and leaves, with seeds, pods and storage organs (tubers, rhizomes, corms) also important. Chimps and baboons share considerable overlap with humans in food items, implying similar overlap and competition between these forms and early hominids when they are found in the same area. No instances are known of chimp (or other pongid) overlap, but fossil baboons and similar forms are known from many early hominid sites.

Peters and McGuire (1981) made a detailed study of the Makapansgat area, to reconstruct the plant food niche of *A. africanus* on the assumption that present flora is similar to that of member 3 times. During the wet season (summer) a wide range of plant foods is available, but during the dry, winter months hard nuts and berries are the most frequent potential food items. While *A. africanus* jaws may have been able to process many of these, these authors consider that the hardest of them, which hominids would need to exploit to survive in that environment, must have been prepared artificially. No definite tools are associated with *A. africanus*, but McGuire (1980) reported quartzite pebbles at Makapansgat with fractures consistent with being used as hammers to break open food objects; such tool-using activity is known for chimpanzees and so is not improbable for *Australopithecus* also. Peters and McGuire suggested that the highland areas around Makapansgat were mainly occupied during the wet season, and that during the winter months occupation was concentrated in lower lying areas with a wider range of plant foods.

Several workers have attempted to infer diet from the study of dental wear and morphology. Wear in South African fossils is spectacular, amounting to virtual destruction of the entire tooth crowns in older individuals, but before that stage is reached some interesting patterns emerge. Robinson (1956) suggested that more frequent chipping of *A. robustus* teeth indicated a greater vegetable content contaminated with dirt and grit, but Wallace (1973) could find no significant difference in the instance of chipping between *A. africanus* and *A. robustus*. He did, however, note differences in the development of wear, which he thought indicated improved crushing and grinding in *robustus*.

Grine (1981) further studied wear patterns in milk teeth by electron-microscopy, noting differences in dental cusp morphology and wear pattern between *A. africanus* and *A. robustus*, which he interpreted as reflecting a greater degree of shearing in the former and more grinding and puncture crushing in the latter. In the light of the palaeoenvironmental evidence, Grine concluded that the robusts were adapted for increased crushing and grinding activity, and were eating harder, more fibrous, more resistant (and perhaps smaller) food objects than *A. africanus*.

Later work (Kay and Grine, 1988) extended microwear analysis to the permanent dentition (M_2) of *A. africanus* and *robustus* and compared results with modern primates. Overall, the patterns of scratches, pitting and wear facets suggest frugivory in both species, but with differences between them. *A. africanus* wear resembles that of modern primates feeding on mature, fleshy fruit, swallowing pulp and seeds whole. *A. robustus* wear is like that of forms eating seeds encased in tough outer coats, crushing the seeds between their cheek teeth. The greater frequency of occlusal pitting in *A. robustus* points to a similar conclusion, since it resembles those forms feeding on hard foods such as date palm seeds and palm nuts.

Walker (1981), focusing on the East African fossils, also studied patterns of tooth wear and compared them with modern analogues. He showed in studies of hyraxes that browsers and grazers could be distinguished, but that wear patterns change rapidly and so reflect activity for only a short period before death. Walker concluded that the microwear patterns of *A. boisei* excluded both grazing and browsing and is most consistent with frugivory, as exemplified by modern chimpanzees and mandrills, a rather unexpected result given the general palaeoenvironmental contexts of *A. boisei* (but see p. 105).

Walker further noted that biomechanical reconstruction of *A. boisei* and human jaw muscle actions are very similar and that, given the greater area of the *A. boisei* dentition, occlusal pressures (as opposed to forces) were probably similar. The diet may not have been especially hard, but one which required food in quantity, possibly because it contained much indigestible material. This view (which is also that of Isaac, 1981) is similar to Wallace (1975) and contrary to the conclusion of Grine (1981) and Kay and Grine (1988) of different diets for the South African robusts.

There is thus considerable uncertainty about early hominid diets, but this is not surprising given the limitations of the data and the eclectic behaviour of modern primate species in this respect. Current interpretations are agreed in laying much greater emphasis on plant food items, and the recognition that tree products as well as terrestrial vegetation would have been utilised, so paralleling current views on early hominid bipedalism as a locomotor pattern involving arboreal clambering as well as terrestrial activity. The most pressing need is for more sites sampling a wider range of environments, so that some of the present uncertainties can be resolved.

CHAPTER SIX

PLIO-PLEISTOCENE HOMINIDS

6.1 Introduction

Hominid remains from the final Pliocene and basal Pleistocene (*c.* 2.0–1.4 mya) exhibit unparalleled diversity. *Australopithecus robustus* and *A. boisei*, with their distinctive morphology (see chapter 5) are known from South and East Africa, but there are other hominid fossils from both regions that exhibit contrasting features of the skull and dentition. Some appear to have been fairly large bodied, while others are more diminutive, and no bigger than small mid-Pliocene specimens such as 'Lucy' or Sts 14. However, all contrast in cranial and dental features with australopithecine species and show undoubted resemblances to later human fossils. These 'non-robust' Plio-Pleistocene specimens are therefore considered to be the earliest members of the genus *Homo*, and are frequently included within the species *H. habilis*.

At around 1.8 mya or slightly earlier, specimens of *H. erectus* grade appear (see chapter 7), so that within a span of 0.3–0.5 my (< 2.0–1.5 mya) at least three, and possibly more, hominid morphologies are in evidence (Leakey, 1976a). It is not surprising that such diversity has generated a range of phyletic and adaptive interpretations. Early *Homo* fossils are few and fragmentary, and yet their various groupings and interpretations critically determine how early human evolution is viewed. There is as yet no consensus on this and the debate is more complex than for most other aspects of palaeoanthropology. In this chapter, the more important fossils are therefore described individually, followed by their possible morphological and taxonomic groupings, and the evolutionary schemes based on these. Finally there is a survey of adaptive and behavioural reconstructions based on this important material.

Early *Homo* fossils are known from several localities in East and South Africa, but the bulk of the evidence comes from the critical sites of

Table 6.1 Some important Plio-Pleistocene hominid sites.

Site	Age (mya)	Hominid species
Olduvai Gorge, Tanzania Beds I and Lower II	1.5–1.8	*A. boisei, H. habilis*
Koobi Fora, Kenya (East Turkana)	1.4–2.0	*A. boisei, H. rudolfensis, H. habilis, H. ergaster/erectus.*
West Turkana, Kenya	1.4–2.0	*A. boisei, H. erectus*
Omo, Ethiopia (Member G)	*c.* 1.9	*A. boisei, H. habilis*
Peninj, Tanzania (Lake Natron)	1.4	*A. boisei*
Lake Baringo, Kenya	1.4	*A. boisei*
Sterkfontein, South Africa (Member 5)	?1.5–2.0	*H. habilis*
Swartkrans, South Africa (Member 1, 2)	?1.5–2.0	*A. robustus, Homo* sp.

Olduvai Gorge, Tanzania, and around Lake Turkana, Kenya. Virtually all specimens were recovered through fieldwork from 1960 onwards.

6.2 East Africa

6.2.1 *Olduvai Gorge*

Olduvai Gorge forms part of the western Rift Valley in northern Tanzania, cutting into the upland Serengeti Plain. It consists of two arms, the main and side gorges, about 14 and 8 miles long respectively. Recognising its potential as a hominid site, Louis Leakey began excavating in the gorge in the 1930s and recovered much faunal and archaeological evidence before the first hominid fossils were found from 1959 onwards.

Overlying the basement rocks are deposits of late Pliocene/Quaternary age, generally some 45–80 m deep but in places reaching up to 110 m (Hay, 1976). They are divided into five main units (Beds I–V); numerous tuffs, especially in Beds I and II have proved invaluable for correlating stratigraphy along the gorge and for absolute dating. Six tuffs (I_A–I_F) are recognised in Bed I and four (II_A–II_D) together with the Lemuta member (Aeolian tuff complex), an important marker within Bed II. Potassium–argon dates on these indicate that Bed I covers the period 1.9/1.8–1.7 mya; Bed II 1.7 to *c.* 1.2 mya; Bed III 1.2–0.8 mya; and Bed IV *c.* 0.8–0.6 mya.

Important hominid localities lie along the entire main gorge and part of the side gorge, with a particular concentration around where the two meet,

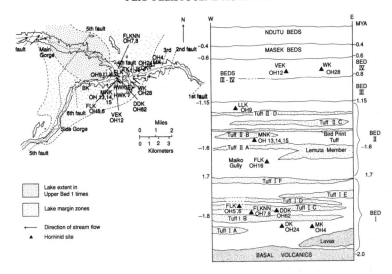

Figure 6.1 Olduvai Gorge. Left: map of gorge with main hominid sites. Right: Schematic composite section of the main gorge with some important specimens indicated (data from Hay, 1976).

some 6 miles from the main gorge's eastern end. The sites are mainly within Beds I and II, with some from Bed IV; Bed III is largely non-fossiliferous, and the area was apparently unoccupied by hominids at that time.

During Bed I times a shallow alkaline lake existed in the gorge, varying in extent from 7–25 km. Its salinity indicates a relatively dry climate overall, but with some fluctuations: it was wetter before tuff I_D (1.7 mya), drier thereafter. Early hominid sites in Bed I are concentrated around the former lake edge, especially its south-east margin near the junction of the main and side gorges, where freshwater streams fed the alkaline lake. Plant, vertebrate and invertebrate fossils all indicate damp, swampy, marshy conditions in this area, which was intermittently flooded by the lake during the earlier part of the sequence.

Lake salinity was appreciably greater and the climate more seasonal during Bed II. There is evidence of torrential rain at times, but soil salinity and wind-blown deposits show vegetation cover to have been insufficient to prevent wind erosion, and point to significant aridity. Faulting around the time of the Lemuta member (mid-Bed II) led to increased drainage and reduced the lake to about one-third of its former size. Significant faunal change ensued, with reduced numbers of swamp-dwelling species and greatly increased numbers of open-grazing forms.

In 1960 Leakey described remains of an immature individual (OH 7) from Bed I deposits at site FLKNN (Leakey 1960; 1961a,b). They included a mandible with dentition, an upper molar, two parietal bones with other cranial fragments and a group of hand bones, some of which were later shown to be non-hominid. From the same site Leakey also recovered much of a hominid foot (OH 8) and a clavicle (OH 48), which may belong to the same individual as OH 7. The cranium and teeth presented obvious contrasts with the OH 5 (Zinj) fossil found the previous year, and with other australopithecines.

The OH 7 parietal bones are thin and, reflecting their immaturity, lack any crests, tori or pronounced muscle markings. Although incomplete, they preserve parts of the mid-vault suture so that reconstruction of the biparietal arch is possible, and from that total endocranial volume can be estimated as $c.$ 645 cm^3. Allowing for some further growth, Tobias (1991) estimates an 'adult' capacity of 674 cm^3. The mandible lacks the lower border, which has eroded away, but it was clearly strongly built. The anterior teeth are relatively large and the canine bears a small but distinct wear facet on its tip from the upper tooth, which indicates slight canine overlap. Cheek teeth are within the range of australopithecine dental dimensions but the premolars are particularly narrow; the third molars had not erupted at the time of death, equivalent to $c.$ 11–12 years on a modern human standard.

The hand bones are powerfully built with strong, slightly curved finger phalanges and prominent insertions for the flexor muscles. The lower part of the thumb is missing, but its terminal phalanx was short and broad, and was opposable, for the trapezium at its base possesses a characteristic human saddle joint. The foot bones are small compared with modern humans, but reveal full bipedal adaptations in the form of well-developed arches and an adducted big toe. The FLKNN clavicle is stoutly built but otherwise like that of modern man.

In 1963 the Leakeys discovered further hominid remains (OH 13, or 'Cinderella') at the nearby MNK site in rather younger deposits from lower Bed II. The material includes an incomplete, lightly built cranial vault with parts of the frontal, parietal, temporals and occipital, much of the palate with the cheek teeth, a well-preserved mandible with complete dentition, and numerous other fragments. Although not fully mature—the third molars are just coming into occlusion and are unworn—the individual was older at death than OH 7. Cranial capacity (673 cm^3) is virtually identical with the estimate for OH 7, but the jaw and teeth are much smaller and more lightly built. As the nickname suggests, the specimen is generally regarded as female.

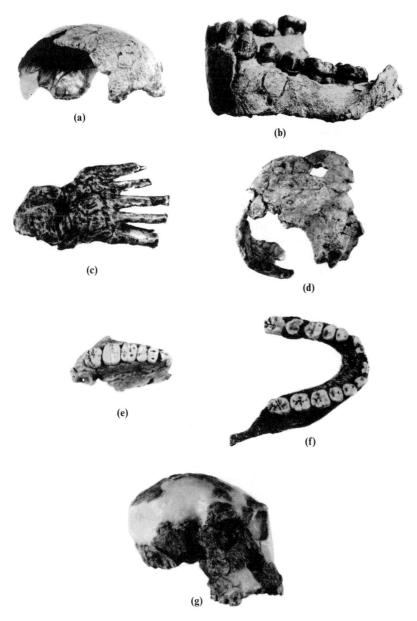

Figure 6.2 Olduvai *Homo habilis* (a) OH 7 vault fragments (b) OH 7 mandible (c) OH 8 foot. (d) OH 13 vault (e) OH 13 maxilla (f) OH 13 mandible (g) OH 24.

The Leakeys also recovered fragments of a hominid cranium and dentition (OH 16, 'George') from site FLK II Maiko Gully. The specimen, washed out by heavy rain, had then been trampled and broken up by Masai cattle. Its large teeth, well-developed brow ridge, strong temporal line and nuchal markings all suggest a male individual. The vault, reconstructed from over 100 fragments, has a capacity of 638 cm³ (Tobias, 1991).

In the same year Leakey et al. (1964) provided a revised diagnosis of the genus *Homo* and defined a new species, *H. habilis*, on the basis of the Olduvai material, with OH 7 as the type specimen, the FLKNN hand and foot bones and clavicle and the OH 13 cranium, as well as various other bits and pieces as paratypes; OH 16 and the fragmentary ramains of another juvenile (OH 14) from MNK II were referred to the species but not included in the type sample. For further discussion of *H. habilis* see below.

In 1969 Mary Leakey, working at the DK site, recovered a remarkably complete cranium (OH 24, 'Twiggy') from deposits low down in Bed I, making the specimen the oldest of the Olduvai hominids. Most of the skull was preserved, including the greater part of the vault, face and palate and some of the cranial base, but it was much crushed and distorted; even after reconstruction by Clark some distortion remains, producing a flattened and elongated vault. The specimen is small and lightly built (cranial capacity just under 600 cm³) with a concave, prognathous face in which some detect australopithecine-like features. The palate and teeth are small.

The next major discovery at Olduvai was in 1986 when White discovered a partial hominid skeleton (OH 62) from Bed I deposits at a new site, DDH, close to the original FLK ('Zinj') site and of about the same age (1.75–1.85 my) (Johanson et al., 1987). The trunk is missing, but parts of the skull, right arm and both legs are preserved. The individual was small, but the forelimb long and relatively robust while the hindlimb was short and the femur lighter even than that of 'Lucy' (AL 288A). The skull remains include part of the braincase, mandible and dentition, palate and lower face, and show resemblances to OH 24 and the South African fossils SK 847 and especially STW 53 (see below). The find also provides information on bodily proportions in Plio-Pleistocene hominids, and reveals that some individuals were short and had powerfully proportioned forelimbs and relatively short hindlimbs reminiscent of 'Lucy', more than c. 1 my earlier.

6.2.2   *Lake Turkana*

Lake Turkana (formerly Lake Rudolf) lies within the mid-Pliocene Turkana basin of north Kenya, fed by the Omo at its northern end and by several other rivers from east and west. The lake's extent (currently 180 × 30 miles) has fluctuated markedly over the last 3–4 my following climatic changes, earth movements and volcanic events. As a result, fluvial, deltaic and lacustrine sediments together with multiple volcanic tuffs form deposits > 300 m thick within the basin. The present landscape is largely one of semi-arid, hilly 'bad lands' subject to wind erosion, and dissected by gullies from seasonal streams and flash floods. The area is richly fossiliferous, with hominids and artefacts scattered over a wide area around the lake. Most finds are from the eastern side where they cover > 1300 square miles from Ileret in the north, through the area behind the base camp at Koobi Fora to Allia Bay in the south. Richard Leakey and co-workers have excavated here since 1968; in recent years they have made important finds in the west Turkana region.

Faulting and other factors made correlation between scattered localities difficult and bedevilled attempts at an overall stratigraphic framework. Numerous tuffs, of which the most important are the Tulu Bor, Ninikaa, KBS, Okote and Chari tuffs, eventually provided a basis for correlation and dating, but some initial mismatches occurred, making for further difficulties. An early and influential scheme was that of Bowen and Vondra (1973), but it is only recently that a more comprehensive framework has been established (Brown and Fiebel, 1986) that clears up inconsistencies

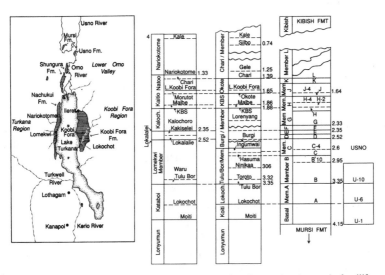

Figure 6.3 Lower Omo and Turkana basin. Left: regional map showing main fossiliferous areas. Right: correlation and dates of main deposits in (from left to right) West Turkana region, Koobi Fora (East Rudolf) region, Main Omo (Shungura) Formation and Usno Formation (data from Feibel *et al.*, 1989).

Table 6.2   Stratigraphic schemes for the Turkana basin. In Bowen and Vondra's scheme the 'Suregei tuff complex' (STC) defined the base of the Koobi Fora Formation. Due to miscorrelation both the 'STC' and the 'Tulu Bor tuff' as then identified included what are now known to be several different horizons. Some 'STC' exposures derive from the upper Lonyumun Member, but others are much younger and correlate with the Burgi tuff. Some exposures of the KBS and Lorenyang tuffs were mistakenly identified as the Tulu Bor tuff which is, in fact, much older than these.

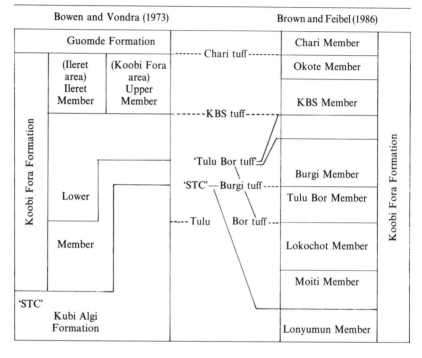

and relates the Turkana sequence to those of Omo, Hadar and elsewhere. Table 6.2 provides a correlation of the 1973 and 1986 schemes; the former's terminology is the basis for many important publications on the Turkana finds.

Most of the hominids are from the Koobi Fora formation above the Lorenyang tuff (c. 2 mya) and below the Chari tuff (1.39 mya). Several important finds straddle the KBS tuff, which is an important marker, separating the upper part of the Burgi member from the overlying KBS member. Initial ^{40}Ar–^{39}Ar dates gave an age of 2.61 my for the KBS tuff, later revised down to 2.43 my, implying hominids beneath the tuff were around 3.0–2.7 my old. However, faunal incompatibilities and other anomalies suggested that these dates were too old, and subsequent potassium–argon dating indicated the KBS tuff age to be 1.88 my old, a figure confirmed by other evidence. The Okote tuff (1.64 my old) in turn separates the KBS and Okote members, with the latter bounded by the Chari tuff. The earliest hominids from the Upper Burgi member below the KBS tuff are thus c. 2 my old, and the latest Plio-Pleistocene specimens just below the Chari tuff c. 1.4 my old.

Geological fieldwork has revealed a variety of palaeo habitats within the basin: lakeside, alluvial delta and flood plain, and riverine localities. The overall environment, whilst open, was wetter than the semi-arid landscape of today, and was probably open or wooded savannah. There was dense vegetation along the lake margin, except where periodic flooding produced swamp and mud flats seasonally covered with grass, and gallery forest fringed the major rivers. Their extents changed over time, but throughout the sequence the Turkana region contained a variety of habitats (see sections 6.8 and 6.9).

A large number of hominid specimens are known from localities within the Koobi Fora formation, including crania, mandibles and postcranial remains. The cranial material is especially varied; besides large *A. boisei* individuals (see chapter 5) there is a wide range of other material. Some specimens are small and lightly built: KNM ER 1813, which is the most complete, has a small thin-walled vault (500 cm^3) and only slight muscular markings. The supraorbital torus is only moderately developed, and the mid-face is flat with salient nasal bones and limited prognathism. The cheek

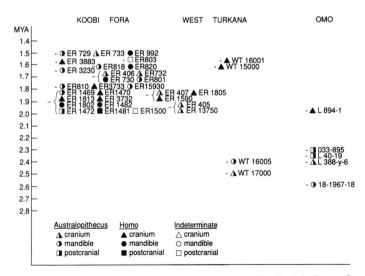

Figure 6.4 Temporal distribution of some important hominid fossils in the Lower Omo and Turkana basin (data modified from Feibel *et al.*, 1989).

is lightly built and the palate and teeth are small, with the dental arcade more parabolic and the cranial base more flexed than in *Australopithecus*.

Another small specimen is KNM ER 732, a right half-cranium including much of the face and side of the palate, and the frontal, parietal and right temporal bones. The vault bones are thin but the mastoid region is ruggedly built; and there are strong muscle markings. The cheek is deep and slopes downwards and forward in a manner reminiscent of robust australopithecines so that the specimen is widely regarded as a female *A. boisei*.

KNM ER 1470 is a relatively complete cranium with much of the vault and part of the face, including the orbital areas and the maxilla, but lacking tooth crowns. The vault is thin walled and relatively long and ovoid with a capacity of $775 \, cm^3$, reflecting expansion of the frontal region, and transverse expansion of the parietals and occipital compared with smaller brained crania. Both supraorbital torus and nuchal ridge are only moderately developed; the face is long, broad and remarkably flat, with very little prognathism. Although exact positioning of the face and braincase is uncertain because of warping, the specimen is complete enough to indicate deep zygomatic processes, set well forward and sloping downwards.

KNM ER 3732 is a less complete cranial fragment including the left orbital region, torus and upper part of the vault. From its dimensions cranial capacity was again relatively large: the frontal is very broad, and, while sagittally short, this is compensated by very long parietals. KNM ER 1590 is a juvenile, fragmented braincase and associated teeth. It is lightly constructed (in part owing to immaturity) and includes most of the two parietals, part of the frontal and other vault fragments. Again cranial capacity was relatively large, and the adult value would probably have exceeded that of KNM ER 1470.

Another specimen, KNM ER 1805, includes the great part of the skull with braincase, facial fragment and mandible. Endocranial capacity is a modest $600 \, cm^3$ or so, but the vault is ruggedly constructed with medial thickening and strong muscular attachments: the temporal lines are well marked anteriorly, and there is a small but distinct sagittal crest. The extensive nuchal area is broad and flat, continued laterally by a well-developed crest above the mastoids. The palate is large (although further expanded by post-mortem breakage) and the teeth moderately sized. The mandible body is fairly robust, but remarkably shallow given the construction of the rest of the cranium.

There are many fossil mandibles: some exhibit the characteristic *A. boisei* morphology while others are more lightly built (Leakey, 1976b; Wood, 1976). Wood (1978) separated the latter specimens into two groups. In one

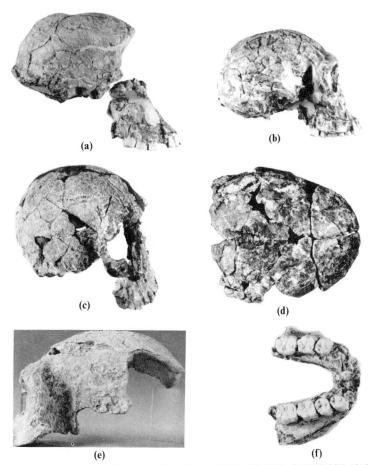

Figure 6.5 Koobi Fora early *Homo*. *H. habilis*: (a) KNM ER 1805; (b) KNM ER 1813. *H. rudolfensis*: (c) KNM ER 1470; (d) KNM ER 1590; (e) KNM ER 3732; (f) KNM ER 1802.

(e.g. KNM ER 1801, 1802) the corpus is relatively broad and stoutly built with some buttressing at the symphysis. Cheek teeth are moderate in size and almost parallel, and the anterior premolar is set obliquely. The other group includes smaller and lighter specimens (e.g. KNM ER 730 and ER 992) with a lightly constructed corpus and less buttressed symphysis. The molar rows are smaller than in 1801/1802 and more divergent, reflecting the more parabolic dental arcade.

H. erectus fossils are also known from sites around the lake. Among the more important, described in more detail in chapter 7, are a beautifully preserved skeleton (WT 15 000) from Nariokotome, West Turkana, dated to

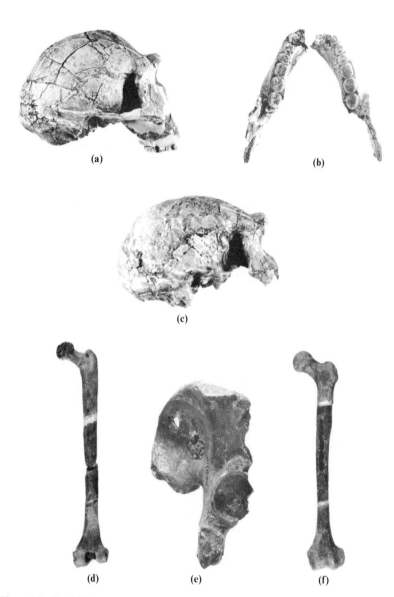

Figure 6.6 Koobi Fora early *Homo. H. ergaster/erectus.* (a) KNM ER 3733; (b) KNM ER 992; (c) KNM ER 3883. Postcranials: (d) KNM ER 1472 femur (*Australopithecus* or *H. habilis*); (e) KNM ER 3228 hip bone (*H. rudolfensis* or *H. ergaster*); (f) KNM ER 1481 femur. (*H. rudolfensis* or *H. ergaster*).

1.6 mya and from Koobi Fora a complete cranium (ER 3733) dated to 1.78 mya. They provide evidence for *H. erectus*-like morphology by at least 1.8 mya and possibly earlier; a hip bone (ER 3228) dated at 1.95 mya is indistinguishable from later *H. erectus* specimens but lacks association with identifying cranial material (Rose, 1984a).

Nariokotome apart, postcranial remains include at least four partial skeletons and numerous fragmentary or complete long bones, especially femora. This material has not been analysed in such detail as the crania, but the larger specimens, including a partial skeleton (ER 1808) much affected by pathology, are usually assumed to be of *H. erectus* grade.

Early *Homo* is represented by two smaller part-skeletons, ER 1500 and ER 3735, both from the Upper Burgi member and dated at *c.* 1.9 mya. Although fragmented and abraded, ER 3735 includes parts of the head, arm and leg, and provides significant information (Leakey and Walker, 1985; Leakey *et al.*, 1989). The skull fragments are small and lightly built, similar to ER 1813. Despite this, the forelimb was similar in size to that of a male chimpanzee, with indications of powerful shoulder muscles and strong flexion at elbow and digits that suggest climbing ability. It is, in fact, much larger than Lucy's forelimb, although sacrum and hindlimb are more nearly equal in the two fossils. A similar, though less extreme, pattern is seen in the smaller OH 62 when compared with Lucy. Leakey *et al.* (1989) suggest ER 3735 and OH 62 represent male and female, and estimate their weights at *c.* 40 kg and 25 kg respectively, similar to the body size dimorphism of chimpanzees.

6.2.3   *Omo*

Parts of a cranium and cheek teeth (L894-1) are known from a layer of silty sand at the edge of a shallow alkaline lake in upper member G of the Shunguru formation (*c.* 1.9 mya). The specimen is much fragmented, but palatal size can be estimated from the cheek teeth, and this, together with parts of both cheek bones, shows the face to have been relatively flat. Similarly, curvature of the few vault fragments suggests a braincase larger than in *Australopithecus*. The cheek teeth, which are fairly worn, are similar in size to those of OH 13 and OH 24 (Boaz and Howell, 1977).

6.3   South Africa: Transvaal sites

Material assigned to early *Homo* is also known from South Africa. Back in the 1940s Broom and Robinson, excavating at Swartkrans, had recognised some hominid dental and jaw remains as non-australopithecine and given them the name 'Telanthropus'. One mandible (SK 15) is probably *H. erectus* (see chapter 7); other 'Telanthropus' specimens included a second, less complete, mandible fragment (SK 45) and a palate (SK 80) from member I. In the late 1960s Clarke recognised that the palatal fragment could be united with parts of a face and cranial base (SK 846b; SK 847) which Broom and Robinson had considered to be *A. robustus*. The composite reconstructed specimen (now classed as SK 847) includes the greater part of the face, the brow region and parts of the side and cranial base of decidedly non-australopithecine form with salient nasal bones and lightly built zygomatic region, so that the specimen is generally included in early *Homo* (Clarke and Howell, 1972). The SK 45 mandible fragment almost certainly represents the same individual. Later Clarke identified other Swartkrans material including an immature cranium (SK 23) and jaw fragments, as *Homo* rather than *A. robustus*.

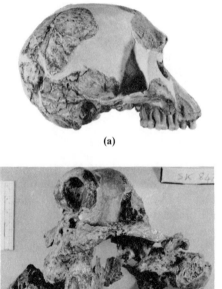

(a)

(b)

Figure 6.7   South African early *Homo*. (a) Stw 53 (Sterkfontein Member 5); (b) SK 847 (Swartkrans Member 1).

In 1976 Tobias and Hughes recovered a partial cranium (Stw 53) from the member 5 breccia at Sterkfontein, considered to be considerably later than the *A. africanus* member 4 breccia. Fragments of the greater part of the skull were recovered, including much of the vault, face and palate, with most of the cheek teeth in place (Tobias, 1978). The nasal bones project, the mid-face is relatively flat and the brow ridge relatively strongly developed, especially medially. Overall, the specimen closely resembles some of the Olduvai crania, particularly OH 24 and, in its palate, OH 62, but has stronger muscle markings, especially for the temporals, than any of the Olduvai specimens.

6.4 Systematic interpretations

6.4.1 Homo habilis

In their announcement of *H. habilis*, Leakey *et al.* (1964) provided a revised diagnosis for the genus *Homo* as well as the species *H. habilis*, stressing the need to take account of the overall features of the material, and not rely on just one or two characters. They focused on postcranial adaptations of truncal erectness, bipedal locomotion and precise manipulation, as well as cranial features, where they revised the minimum threshold for cranial capacity down to $600\,cm^3$ or thereabouts rather than the $700–750\,cm^3$ of previous definitions. Individual specimens of early *Homo* may thus overlap with *Australopithecus*, but the mean value in *Homo* is greater both absolutely and in relation to body size. Muscle markings may be weak or strongly developed and the face flat or projecting, but not dished as in *Australopithecus*. In *Homo* the teeth are set in an even dental curve and the molars are smaller than in *Australopithecus*, while incisors and canines are large relative to the cheek teeth. The teeth generally, and especially the cheek teeth, are narrow compared with *Australopithecus*.

The species *H. habilis* possesses most of the features noted above: relatively large incisors compared with *Australopithecus* or *H. erectus*, and canines large in proportion to the narrow premolars; a tendency for narrowing and elongation of all teeth, but especially marked in the lower premolars and molars; sagittal curvature of the cranial vault, slight to moderate in the parietal and slight in the occipital regions. The hand bears robust, curved and strongly marked digits compared with later *Homo* species.

Both the revised genus and new species were strongly criticised at the time, for many argued that the Olduvai specimens could be accommodated within existing taxa: the Bed I specimens in *Australopithecus* and the Bed II material in *H. erectus*. However, discoveries over the last 25 years and the much larger fossil sample now available confirm Leakey *et al.*'s (1964) recognition of basal Pleistocene hominids that were neither australopithecines nor like later fossils.

Tobias (1991) has now provided a more detailed description of *H. habilis* based on the Olduvai fossils. Brain size, averaging 640 cm^3, is 20–45% greater than in *Australopithecus*, but well below that of *H. erectus*, while encephalisation measures also give intermediate values. Moreover, the general proportions of the brain are much more like those of later *Homo* (see below) than are those of *Australopithecus*. Greater brain size is reflected in a broader, higher, thin-walled braincase—there is little increase in length.

H. habilis had relatively large anterior teeth (as far back as the first molars) but reduced cheek teeth, especially in breadth, and a tendency towards root fusion. Back molars, especially in the palate, are much smaller, the enamel thinner and the teeth less worn than in *Australopithecus*. Although projecting, face and jaws are more lightly constructed and the masticatory muscles less marked than in *Australopithecus*. The lighter face and expanded braincase make for a head which is better balanced on the vertebral column, although the degree of basicranial flexion cannot be determined since the only Olduvai specimen preserving the base (OH 24) is crushed and distorted in this region.

6.4.2   *Diversity in early* Homo

Significant variation within early *Homo* has been recognised ever since the formal announcement of *H. habilis*. Tobias and von Koenigswald (1964) noted differences between the Olduvai Bed I and Bed II fossils, considering the latter to be more advanced, while Leakey (1966) viewed contrasts between the Bed II fossils (OH 13 and OH 16) as divergence towards *H. sapiens* and *H. erectus* respectively. Cranial and dental contrasts between Bed I OH 7 and 24 were usually interpreted as sexual dimorphism, although some OH 24 facial features are australopithecine-like (see above). Discoveries at Lake Turkana in the early to mid 1970s diverted attention away from Olduvai. The Koobi Fora finds greatly increased the hominid sample but their morphological range made for interpretive complexity. Apart from the relatively complete crania of *A. boisei* (e.g. ER 406) and *H. erectus* (e.g. ER 3733), specimens were usually assigned to *Australopithecus* or *Homo* but not to species within those genera, while others were simply placed in suspense. This caution did not prevent informal labelling by some workers and, on occasion, formal attribution.

For example, Groves and Mazak (1975) noted that some specimens from above the KBS tuff differed from Olduvai *H. habilis* in dental proportions and morphology, having smaller cheek teeth and fewer premolar roots. They therefore created a new species, *H. ergaster* ('worker man'), with ER 992 as the type. Unfortunately they failed to distinguish adequately between their new species and *H. erectus*, and it has been strongly argued that the type mandible is, in fact, *H. erectus* (Leakey and Walker, 1989). *H. ergaster* has therefore not been widely recognised, although it has recently been independently revived by Groves (1989) and Wood (1991; 1992a,b) (see below and chapter 7).

Other workers also considered more than one species (quite apart from *A. boisei* and *H. erectus*) to be represented at Turkana but there was little or no agreement as to how the samples should be partitioned. Walker and Leakey (1978) argued for three hominid clusters on either side of the KBS tuff: below were robust and gracile australopithecines and early *Homo*, while above were robusts, graciles, and *H. erectus*. Wood (1978) grouped KNM ER 1470, 1590 and 3732 on vault proportions and capacity, with ER 1805 and 1813 another cluster, and KNM ER 3733 a third distinct form. Howell (1978) combined ER 1470, 1590, 3732 and, by contrast with the above, 1813 as *H. habilis*, whereas ER 1805 was associated with ER 3733 and the mandibles 730 and 992 as *H. erectus*.

Later studies pointed up additional contrasts within the Koobi Fora sample. For example, Falk (1983) separated KNM ER 1470 and 1805 on differences in their cerebral circulation patterns, while Wood (1985) noted contrasts in the braincase and the face of KNM ER 1813 and 1470. Facial proportions were further discussed by Bilsborough and Wood (1988) and neurocranial features considered by Lieberman *et al.* (1988), who argued the improbability of the cranial capacities of these two specimens being drawn from a single population since this would require a greater degree of the sexual dimorphism than that of the modern gorilla. However, the basis of this argument has been severely challenged (see below).

From the mid-1980s several workers provided broader surveys, extending their comparisons beyond the Turkana basin. Among the earliest was Stringer (1986) who reviewed early *Homo* from Olduvai and Koobi Fora, considering variation in endocranial capacity, dental dimensions, facial projection and the proportions of the anterior vault. Because of the marked variation in these characters compared with modern forms, he concluded that two species are represented: a larger

Table 6.3 Some recent systematic schemes for early *Homo* cranial and mandibular specimens.

	Olduvai OH 7, 24	OH 13, 16	Sterkfontein Stw 53	Swartkrans SK 847	1470, 1590, 3732	Koobi Fora 1482	1805, 1813	3733, 3383
Stringer (1986)	*H. habilis, Homo* sp.				—*H. habilis*—(+1802)		—*Homo* sp. (+992)	
Chamberlain (1987)	—*H. habilis*——————————— *Homo* sp. ———————————							*H. erectus*
Groves (1989)	—*H. habilis*———————			*Homo* sp. *aff. erectus*	*H. rudolfensis* (+1802)	*H. aethiopicus*	*H. ergaster* (+992)	*Homo* sp. *aff. erectus*
Wood (1991)	—*H. habilis*———————			*H. erectus*	*H. rudolfensis* (+1801, 1802)		—*H. habilis*	*H. ergaster* (+992)
Tobias (1991)	—*H. habilis*————————————————— *H. habilis* —————————————			*H. erectus*				*H. erectus*

bodied form including KNM ER 1470, 1590, 3732, 1802 (mandible) with postcrania KNM ER 1481 (femur) and 3228 (hip bone), and from Olduvai OH 7 and OH 24, and a smaller bodied form including KNM ER 1813, 1805, 992 (mandible) and Olduvai hominids OH 13 and 16. This arrangement thus divides Olduvai *H. habilis* as well as the Koobi Fora sample. Stringer also suggested that the larger bodied form (which, since it includes OH 7, would be called *H. habilis*) is the morphologically more primitive of the two species.

Chamberlain (1987) and Chamberlain and Wood (1987) compared early *Homo* using techniques that allowed adjustment for size-related differences to estimate shape contrasts that were independent of size. Size-related and size-independent variation in extant primates were identified and the fossil forms assessed against these yardsticks. The early *Homo* sample displayed both size contrasts and size-adjusted shape differences well in excess of those found in sexually dimorphic primate species, and it was concluded that the fossils represented at least two species, *H. habilis* (the Olduvai sample) and an unnamed form, *Homo* sp. (ER 1470, ER 1805, ER 1813, SK 847, Stw 53). Each group contains both large and small specimens. Further contrasts were that *H. habilis* was less variable, had thinner vault bones, shorter parietals, smaller jaws relative to braincase size and a more U-shaped dental arch. Its incisors were relatively larger, the cheek teeth elongated and the lower premolar roots reduced compared with *Homo* sp.

Groves (1989) has reconsidered early *Homo* systematics as part of a comprehensive and challenging survey of primate and human evolution. He confines *H. habilis* to the Olduvai and Stw 53 fossils and in addition identifies 4 (!) species of early *Homo* at Turkana. They are: a primitive, poorly known form, *H. aethiopicus*, based on the V-shaped ER 1482 mandible, together with Omo fossils 338 y-6 (juvenile cranium) and mandible 18-1967-18 viewed by most workers as early East African robusts (*A. aethiopicus* or *A. boisei*, see chapter 5). The crania ER 1470, 1590, 3732 and the 1802 mandible represent *H. rudolfensis*, identified by flat nasals, broad, flat mid-faces and upper molars with Caribelli's complex (accessory cusps).

H. ergaster is based on the ER 992 mandible which Groves argues is not *H. erectus* (see above), associating it instead with ER 1805 and 1813. He therefore sees *H. ergaster* as a relatively small-brained form, contrasting with *H. rudolfensis* in its arched nasal bones and broad upper face. The *ergaster* vault is relatively flat and the premolars broad, in contrast to the more curved cranium and narrow premolars of *H. habilis*. Groves views ER 3733 and 3883 (with which he associates SK 847) as *erectus*-like, but representing a more primitive, as yet unnamed species, that closely resembles the likely common ancestor of *H. erectus* and *H. sapiens* (see chapter 7).

Wood has recently provided full descriptions of the Koobi Fora hominids and formalised his analysis of their diversity (Wood, 1991). Study of modern primates (Wood *et al.*, 1991) has provided further details of the range and pattern of primate sexual dimorphism, and allowed separation of traits indicating intraspecific variability from those pointing to interspecific differences. Examples of the former are measures of facial breadth, canine dimensions and cheek tooth breadths; the latter include cheek tooth lengths, vault breadths, P_4 crown morphology and premolar root form. These findings provide the context for investigating early hominid variation. Like most other workers, Wood views Olduvai *H. habilis* as homogeneous but considers the Koobi Fora sample to contain several species. *H. habilis* is one of these and is represented by crania ER 1805, 1813 and fragmentary mandibles 1501 and 1502.

A second species, *H. rudolfensis*, includes crania ER 1470, 1590, 3732, 3891 and mandibles 1482, 1483, 1801 and 1802. Compared with *H. habilis* it displays a larger cranial vault with relatively shorter occipital, a flat face with the mid-region broader than the upper face, flat nasal bones, anteriorly inferiorly sloping zygomatic processes and a large palate but it lacks a distinct supraorbital torus and nasal sill. *H. rudolfensis* mandibles show greater external relief and a sharper, everted base, while the mandibular fossa on the skull base is shallow rather than deep. In the upper jaw the anterior teeth are large and the premolars 3-rooted; in the mandible the cheek teeth have broad crowns, the premolars twin roots and M_3 is large, whereas in *H. habilis* the mandibular cheek teeth are narrow, the premolars mainly single-rooted and the rear molars reduced.

In addition, Wood, like Groves, recognises a Turkana species of broadly *H. erectus* grade, but more primitive than Asian *H. erectus*. It includes the crania 3733, 3883 and several mandibles, 730, 820 and 992. Because of this last it takes Groves and Mazak's (1975) species name *H. ergaster*, so that Wood and Groves identify very different groups, with contrasting features, by this taxon. This *erectus* grade grouping is discussed more fully in chapter 7.

We are therefore currently faced with two broad perspectives on the early *Homo* material: an apparently simple (and therefore attractive) view that includes the entire sample within the original Olduvai species, *H. habilis*, suitably stretched to accommodate them. This view is favoured by, among others, Johanson *et al.* (1987), Skelton *et al.* (1986), Wolpoff (1988) and

Tobias (1987; 1991). It is also, because of its simplicity, the one found in most textbooks. Alternatively, the range of morphological variation is too great for a single species and is better accommodated within at least two, of which *H. habilis* is one. This interpretation is that of Leakey (1976b), Walker and Leakey (1978), Wood (1978; 1985; 1991), Stringer (1986), Chamberlain (1987), Groves (1989) and others.

Both interpretations pose difficulties. The first requires an extraordinary range of internal variation to be posited, and runs the risk of failing to detect significant differences because the grouping is too coarse. Past confusion has certainly resulted from a tendency by some to view *H. habilis* as a dumping ground for specimens not easily assigned to the polarities of *A. boisei* and *H. erectus* so raising suspicion that they are grouped on negative rather than positive criteria. Indeed, it has proved difficult to identify autapomorphies common to all early *Homo* specimens (see below). The detailed descriptions of the Olduvai fossils (Tobias, 1991) and Koobi Fora hominids (Wood, 1991) now available should prevent this state of affairs in future.

From these and other studies it is clear that the material's overall variability is marked. A single species would, in reality, be even more variable since the few specimens available cannot be expected to sample both upper and lower limits of the population range. The case for two early *Homo* species is accordingly a strong one but there are still difficulties of identification, sorting and species demarcation, as the above summaries demonstrate. Furthermore, Miller (1991) criticises the assumptions by Stringer, Wood and others that for cranial capacity a sample coefficient of variation greater than 10 indicates multiple species within the sample. He shows that large samples of single hominoid species frequently have values > 10 for this character and for small samples values > 10 cannot be shown to be significantly different from a value of 10 or below. In other words, no firm conclusion about species number can be drawn on the basis of brain size variability, and arguments based on this character therefore fall away.

Given these difficulties and uncertainties are there any grounds for preferring one scheme or another? Stringer combined Olduvai and Koobi Fora fossils with mixed subsets, but Tobias (1991) sets out a detailed case for the integrity of the Olduvai sample, a view accepted by most workers. Similarly, Chamberlain (1987) and Camberlain and Wood (1987) confined *H. habilis* to Olduvai, distinguishing it from Koobi Fora *Homo* sp. but the ER 3735 skeleton points to *H. habilis* in the latter region. Groves (1989) identified a species (*H. aethiopicus*) largely based upon material otherwise regarded as part of the East African robust clade. For these reasons, as well as his detailed familiarity with the material, it seems most appropriate to follow Wood's (1991) sorting of the specimens, recognising *H. habilis* as well as *H. rudolfensis* at Koobi Fora.

Of course, Miller's strictures also apply to claims for Koobi Fora heterogeneity based on brain size—e.g. between ER 1470 and 1813 (Wood, 1985; Leiberman *et al.*, 1988)—so that separation on this character is non-proven. However, Wood provides an impressive list of other contrasts between the two forms, so the case for distinctiveness is correspondingly strong. Even if the case for a single species was stronger, it would still be better to proceed cautiously by dividing the sample while retaining that possibility: assuming it at the outset will certainly fail to pick up any contrasts that exist.

6.5 *Australopithecus* or *Homo*?

While the first task is to establish the fossil groupings, there are also problems of nomenclature with this material, since it straddles a generic boundary. Leakey *et al.* were severely criticised for failing to adequately differentiate *H. habilis* from *Australopithecus*, but most workers subsequently accepted its inclusion within *Homo*. However, doubts and reservations about particular specimens continued. Tobias wavered for some time about the inclusion of OH 16 in *Homo* because of its large teeth, while others have suggested OH 24 as australopithecine on the basis of small brain size and facial structure.

Walker (1976) presented arguments (the combination of relatively small brain with large face and jaws) for including ER 1813 and even 1470 in *Australopithecus*, while Leakey (1976b) and Walker and Leakey (1978) assigned 1813 to *Australopithecus*. In doing so they weighted brain size as the single most important criterion for generic separation whereas others, impressed by the masticatory features of the fossils, attached importance to facial structure, jaw and tooth proportions, and were content to include ER 1813 in *Homo*. While arguing for placement within *Australopithecus*, Walker and Leakey left specific identity open and were not implying that 1813 is *A. africanus* (or any other previously identified species), any more than those who placed it in *Homo* considered that it necessarily resembled larger brained species such as *H. erectus* or *H. sapiens*. Genera are subjective categories, invariably broader and less tightly drawn than species, with corresponding scope for disagreement at the margins. The main thing is not to confuse debate about cluster recognition—far and away the most important issue—with arguments over names.

6.6   Phylogeny

Phyletic arrangements obviously depend upon whether one, two or more early *Homo* species are recognised, and upon their composition. Tobias (1991) sees 'broad' *H. habilis* evolving from *A. africanus*. He notes derived similarities between *Homo* and robust australopithecines (64 out of 344 cranio dental characters) that are more primitive in *A. africanus* and argues that these indicate a period of monophyly after the available *A. africanus* fossils. Around 2.3–2.5 mya a speciation phase in response to climatic changes split this posited 'derived' *A. africanus* into a robust clade emphasising dental expansion and masticatory power, and a *Homo* lineage characterised by posterior dental reduction and neural expansion. Thereafter evolution within this lineage was anagenetic according to Tobias (1991).

Several cladistic studies have explored the effects of subdividing early *Homo*. Stringer (1987) found both his early *Homo* groups primitive compared with later forms, but the larger bodied form (containing OH 7 and ER 1470) was more primitive than the smaller bodied (OH 13, OH 16, 1813). This finding was confirmed by Chamberlain and Wood (1987) in a separate analysis that removed size contrasts: Stringer's subdivisions had no effect on the remainder of the cladogram compared with treating early *Homo* as a single undivided group.

By contrast, Chamberlain and Wood's subdivisions (Olduvai *H. habilis* and Koobi Fora *Homo* sp.) produced markedly different arrangements. Olduvai *H. habilis* is primitive compared with *H. erectus* and *H. sapiens*, but *Homo* sp. is distant from these and links instead with robust australopithecines. The two *Homo* subsets have contrasting affinities based on differences in face, jaws, teeth and cranial base: *Homo* sp. is 'robust-like' in its reduced prognathism, deep mandible and narrow jaw with short incisor row, combined with a broad upper face and features of the cranial base. *H. habilis* has a narrower upper face, thinner vault and, in its shorter anterior cranial base and expanded occipital, links with later *Homo*.

In Groves' (1989) study his primitive, poorly known *H. aethiopicus* is considered to be the sister group of all other *Homo* species. *H. rudolfensis* is also relatively primitive, whereas *H. habilis* (Olduvai and Stw 53) is strongly autapomorphic and, as such, debarred from direct ancestry of *H. ergaster* and later species. *H. ergaster* itself is less uniquely derived and so is a plausible ancestor for Groves' unnamed *Homo* species of primitive *erectus* grade.

Wood's most recent analysis (1991) shows his *H. habilis* (i.e. including ER

1805 and 1813) and *H. rudolfensis* as sister taxa within a monophyletic *Homo* clade with higher vault, longer cranial base and elongated lower molars. However, this arrangement is only marginally more parsimonious than a polyphyletic one with the two species as parallel developments, a highly relevant consideration in view of the features typifying each species. *H. habilis* is characterised by derived features of the vault, in face, base and mandible, and by reduced cheek teeth. Neurocranial expansion apart, *H. rudolfensis* displays a cranial anatomy that is either primitive or, particularly in face and jaws, derived in ways that mimic robust australopithecus rather than later *Homo*. The apomorph character sets are, in fact, not dramatically different from the earlier findings of Chamberlain and Wood (1987), despite the shifts of ER 1805 and 1813 between the two analyses.

The postcranial evidence further complicates matters. The small-bodied OH 62 and ER 3735 skeletons show forelimb climbing adaptations, if anything even more pronounced than in 'Lucy', associated with small light cranial remains and displaying size contrasts plausibly interpreted as sexual dimorphism. ER 3735 also provides independent, non-cranial evidence of *H. habilis* at Koobi Fora.

Larger Koobi Fora postcrania (e.g. ER 3228 hip bone and ER 1481 femur) are indistinguishable from later *H. erectus* fossils, but they lack association with identifying cranial material. Leakey and Walker (1989) suggest they represent the larger bodied *Homo* species associated with the bigger crania such as ER 1470, 1590 and 3732 (i.e. Wood's *H. rudolfensis*).

Several broad conclusions stand out from this welter of cladistic detail:

(1) Close analysis shows the early *Homo* sample to be so diverse that the 'broad' *H. habilis* concept appears increasingly unrealistic. The issue therefore becomes not whether there is more than one species, but rather how many, their composition, features and interrelationships.

(2) Cranio-dental evidence shows contrasting characteristics for *H. habilis* and *H. rudolfensis*. While of relatively modest brain size, *H. habilis* shows progressive features of cranium, face and jaws, and reduced dentition. *H. rudolfensis*, apart from increased brain size, shows features of face, masticatory system and dentition that parallel robust australopithecines.

(3) Postcranial remains are reliably associated with *H. habilis*, less so with *H. rudolfensis*. However, if the latter's putative associations are confirmed, postcranial contrasts are equally if not more striking than those of skull and dentition. *H. habilis* possesses a remarkably primitive, essentially australopithecine-like postcranium. *H. rudolfensis* evidently has an appreciably more derived locomotor skeleton, like later *Homo*.

(4) Given the combination of progressive skull and dentition with primitive trunk and limbs (*H. habilis*) or derived postcranial skeleton with australopithecine-like cranio-dental features (*H. rudolfensis*), neither species is self-evidently ancestral to later *Homo*. On the contrary, any interpretation needs to invoke multiple parallelisms (above or below the neck, depending on species) to account for these incongruous and bewildering mosaics.

(5) The Olduvai record is slightly later than that at Koobi Fora, so that when *H. habilis* was confined to Olduvai, it was possible to consider geographical variation and/or temporal differences as possible factors underlying the anatomical contrasts: the two morphs might be allopatric subspecies or species, or sequent chronospecies of a lineage (Stringer, 1986; Chamberlain, 1987). Recognition of *H. habilis* at Koobi Fora and clarification of the dating there (Feibel *et al.*, 1989) renders both interpretations untenable, for the two forms are in the same region at the same time.

(6) The scale and pattern of the morphological differences also militate against such interpretations. The contrasts in locomotor systems, masticatory apparatus and dentition are too great to reflect intraspecific variation or differences between closely related, recently divergent species, even accepting the notion of character displacement. Nor do they make sense as sequential shifts within a single hominid clade. They point instead to considerable divergence and, by inference, an appreciable time interval since the last common ancestor, certainly before 2 mya. They also suggest contrasting niches and adaptive strategies, although detailed evidence for these is lacking.

Evidence therefore indicates *H. habilis* and *H. rudolfensis* to represent distinct lineages, synchronic and sympatric in their distribution (Wood, 1992c). It also raises major questions: is the *H. rudolfensis* pattern further evidence of a shared ancestry for robusts and *Homo* before divergence (Skelton *et al.*, 1986; Tobias, 1991)? Does the small, cranially progressive but postcranially primitive *H. habilis* morph represent a persistent, almost relict, *afarensis*-like clade from the mid-Pliocene that finally becomes extinct around the Pleistocene boundary? More associated cranial and postcranial material, especially of the larger bodied *H. rudolfensis*, is urgently needed before the picture can be clarified.

Remains of *H. erectus* grade (*H. ergaster*) (see chapter 7) occur only slightly later, if at all, than the earlier specimens of both *H. habilis* and *H. rudolfensis* so that more than one hominid lineage was characterised by brain enlargement, face, jaw and dental reduction and locomotor developments. The details of this array have major implications for how our own genus is viewed. If there was a short, sharp speciation phase and the lineages shared a common ancestry during the later Pliocene (say *c.* 2.2 + mya) it would still be possible, by suitable redefinition, to identify a monophyletic origin for *Homo*, say Tobias' 'derived *A. africanus*' or something similar. However, it is quite possible that the shared ancestry is more distant (?mid-Pliocene) and that discrete lineages independently crossed the *Homo* threshold, which makes the genus polyphyletic.

Certainly, as currently understood, it cannot be claimed that *Homo* is used other than in a grade sense. While distressing to purists, this approach is reasonable in view of the limited data available and uncertainties over evolutionary relationships. It provides an example of when it is appropriate to 'classify things on the basis of what they look like rather than on what they might evolve into, if by any chance, we happen to have our lineages correctly documented' (Walker, 1976). Arguments for a cladistic definition of the genus, while acceptable in principle, are impracticable at present.

6.7   Summary

Given the complexities associated with the early *Homo* material, a summary is appropriate. Cranial evidence of early *Homo* is now quite extensive, with most material from Olduvai and around Lake Turkana, together with other sites in East and South Africa. Postcranial remains are much less well known, but some (e.g. OH 62 and ER 3735, referred to *H. habilis*) are surprisingly primitive with suggested arboreal traits. Other larger and more 'modern' postcrania are known, especially from Koobi Fora, but it is not clear whether they belong to *H. rudolfensis* or *H. erectus*. Cranial specimens show marked variability: those assigned to *H. habilis* have brain sizes within or little beyond the *Australopithecus* range but face, jaws and teeth smaller than in that genus, and resembling later *Homo*. Bigger crania, identified as *H. rudolfensis*, have capacities well in excess of the australopithecine range, so justifying inclusion in *Homo*, but show some parallel resemblances to *Australopithecus* in tooth and jaw proportions, and certain aspects of the face and skull base. Overall, however, the pattern is distinct from that of any recognised australopithecine species.

Cladistic analysis points to contrasting affinities for *H. habilis* and *H. rudolfensis*, but their phyletic relationships with *H. erectus* grade and later hominids remain obscure. When these are clarified it may well be appropriate to distinguish between a highly derived 'human-like' *Aus-*

tralopithecus species, and a basal member of the *Homo* clade. Until then it seems preferable to refer all specimens to *Homo*, recognising its use as a grade taxon.

6.8   Robust australopithecines

Robust australopithecines, described in chapter 5, are frequent at South and especially East African localities during this period. The majority are dated to between 2 and 1.5 mya; within this interval are the Swartkrans member 1 sample, Olduvai Bed I specimens such as OH 5, the Koobi Fora sample, etc. A few (Chemeron, Baringo, Peninj) are slightly later (*c.* 1.4 mya) The latest in East Africa are teeth from Olduvai Upper Bed II (OH 3,38) at *c.* 1.2 mya but finds after 1.5 mya are much sparser than the preceding 0.5 mya and may reflect dwindling population numbers. Final extinction may be later than the available record, since later East African sites are poorly known. In South Africa the latest representatives are the jaw fragments and teeth in Swartkrans member 3 which, if correctly dated at *c.* 1 mya (Grine, 1988a), makes the sample the latest known anywhere. However, considerable uncertainty attaches to this estimate. There are no appreciable differences in tooth dimensions between the robusts in Swartkrans members 1–3, suggesting a period of stasis in dental characters, although Wolpoff (1988) detects significant changes in East African *A. boisei* over roughly the same period (see below).

It has been suggested (Klein, 1988) that the major climatic change *c.* 0.9 mya may have finally led to robusts' extinction, but this must remain speculative until more sites in the relevant time range are known. Other possible explanations include the decline and extinction of potential robust predators such as sabre tooths and some hyena species, and their replacement by more efficient carnivores. Again, progressive niche expansion of *Homo* groups may have adversely affected robust species, but only long after the appearance of *H. erectus*. These explanations are not, of course, mutually exclusive, and multiple agencies are more likely than single ones, but as yet there is not enough evidence to determine cause(s) with any confidence.

6.9   Environments and habitats

With many localities yielding both robusts and early *Homo* fossils there are indications of at least broad sympatry, and interest has naturally focused on habitat preferences and possible interactions, particularly in East Africa. Behrensmeyer (1978) and Leakey *et al.* (1978) suggested that at Turkana both species exploited lake margin localities, with *A. boisei* also frequent in fluvial deposits, reflecting bush and gallery forest, whereas *Homo* was confined to the lake edge. A later study (Behrensmeyer, 1985) suggested at least partial ecological exclusion, with *A. boisei* in habitats with greater vegetation cover. Similarly, Shipman and Harris (1988) suggested differences in habitat preference, with *A. boisei* in closed/wetter contexts, and early *Homo* more catholic in dry, open and bushy habitats as well as wetter ones.

However, their method of habitat reconstruction has been criticised (see p. 105), while White (1988) has drawn attention to the disturbing effects of depositional and other taphonomic factors, and the need for chronological controls. He points out that at least some of the above contrasts reflect broad changes within the Turkana basin, with lake habitat prevalent in the earlier part of the sequence and river localities more frequent after *c.* 1.8 mya. His refined analysis shows *Homo* to be more frequent than *A. boisei* in early (Upper Burgi) coastal localities but the frequencies to be reversed in the KBS member, so that *A. boisei* is more numerous in the same contexts. Both forms are equally frequent in the highest (Okote) member. The causes of these shifts are unknown: among them may be sampling errors, changing populations numbers and shifts in behavioural preferences.

6.10 Behaviour

Unlike earlier periods, when hominid behaviour is shadowy and inferential, finds around the Plio-Pleistocene boundary provide definite evidence of early hominid activity: sites at Olduvai, Koobi Fora, the Omo Valley and Sterkfontein have all yielded artefacts. Most of these localities have also provided remains of more than one hominid, usually robust australopithecines and one or more forms of early *Homo*. The tools are generally considered to have been made by *Homo*, although this is assumption not fact, and all Plio-Pleistocene hominids appear to have possessed the prerequisite anatomical structures for tool making; it is possible, indeed probable, that *all* hominids at around this period were tool makers.

The earliest artefacts may be those from Hadar, from localities west of, and rather younger than, the main hominid sites. Here surface collections and a test excavation by Harris yielded flakes and small cobbles from which flakes had been struck (Lewin, 1981). The site (a river bank location), together with others in the same area is estimated to be 2.7–2.4 my old.

Other early sites are those in the Omo Valley, Ethiopia; tools at two localities *may* derive from member E (*c.* 2.4 mya), and there are more definite, *in situ* finds from member F (*c.* 2.3 mya) (Howell *et al.*, 1987). The evidence is sparse and consists of localised clusters of sharp quartz flakes struck from small pebbles. Retouch is very rare, and even signs of wear are uncommon. One or two sites also yield bones, but in some instances these are rolled, and elsewhere are absent altogether. The occurrences are restricted in extent and thin in section, implying transient activity not settlement, but some are repeated, suggesting return visits. All occur in river edge contexts, close to tree cover (Howell *et al.*, 1987).

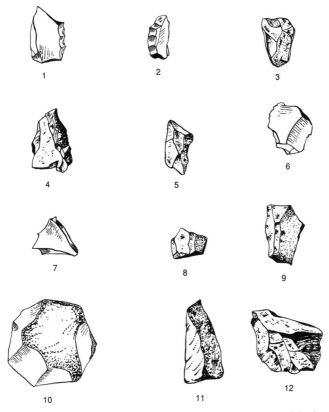

Figure 6.8 Early artifacts from localities in Omo Member F. They are mainly sharp flakes, some showing retouch and with occasional discoid cores (from Merrick and Merrick, 1976).

A similar assemblage is known from the recently excavated site of Senga 5a, Semliki River, Zaire, dated on faunal grounds to Shungora F–G (2.3–2.0 mya). This lakeside site documents hominid occupation of east Central Africa, on the western edge of the Rift Valley, by the late Pliocene (Harris *et al.*, 1987). Rather later occurrences at Koobi Fora are also generally thin and scattered. The exception is Olduvai Gorge, where there is a greater concentration of early sites than elsewhere; those in Bed I/Lower Bed II were located among reeds and marshes at the edge of the saline lake within the gorge.

The earliest tool assemblages are generally referred to as Oldowan or 'pebble tool' industries, reflecting the use of stones with one or more flakes removed to form an irregular sharp edge. These are usually viewed as pounders and choppers, and some show signs of having been used in this way. However, flakes are far more numerous at all sites, and it is likely that many of the 'choppers' are simply the discarded cores used to produce flakes, which experiment shows to be especially effective for piercing and severing hide or disjointing/disarticulating carcasses. At many sites only the later stages of flake removal are represented, implying hominid transport of cores or flakes to and/or from the site for future use (Toth, 1987b). At some locations through-flow was appreciable, implying significant use away from the site and transportation, perhaps in response to the immediate availability of raw material (Schick, 1987). Some artefact clusters may represent 'caches' in the landscape, deposited for future use (Potts, 1984). Transportation on this scale reveals not only selectivity by hominids, but detailed knowledge of resources, and of foresight and planning within a temporal framework.

It has been suggested that such mental skills were selected for by increased meat eating, given the patchy and unpredictable distribution of prey and carcasses across the landscape, but it could just as well have been the other way around. Tool-making skills and knowledge of environmental resources may well have promoted meat eating. In any case, many items, including a range of plant foods, are unpredictably scattered across the savannah environment.

The archaeological record undoubtedly indicates complex behaviours; can anything more be learned from the endocast evidence? This is notoriously difficult to interpret (see chapter 5), but Tobias (1987) has shown that early *Homo* cranial capacities were significantly larger, absolutely and relatively, than those of *Australopithecus*. The early *Homo* average is 640 cm^3, with a range of 510 cm^3 (ER 1813) to 752 cm^3 (ER 1470); for Olduvai *H. habilis* the mean is 645 cm^3, with a range of 594–674 cm^3.

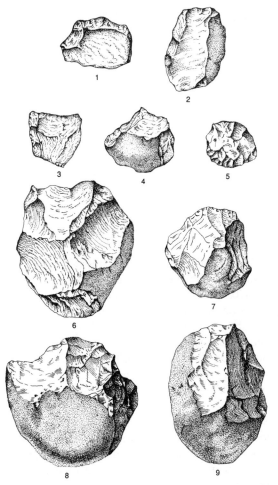

Figure 6.9 Oldowan artifacts, Bed I Olduvai Gorge. Scrapers and retouched flakes (1–5); choppers and cores (6–9) (from Leakey, 1971).

These compare with means of 400 + cm³ to 530 cm³ for the various species of *Australopithecus* (see chapter 5). All measures of encephalisation (chapter 3) show early *Homo* well ahead of *Australopithecus* (Tobias, 1987), with relative EQ values of around 55% of the modern human figure, compared with 30–40% in australopithecine species.

The increased cranial volume is seen in some heightening and more marked broadening of the endocast, especially of the frontal and parietal

Table 6.4 Brain and body size estimates and encephalisation measures in early hominids. Fossil values from McHenry (1988), *H. sapiens* values from Aeillo and Dean (1990).

	Body weight (kg)	Brain size (cm³)	EQ		%EQ	
			Martin	Jerison	Martin	Jerison
A. afarensis	50.6	415	1.87	2.44	30	30
A. africanus	45.5	442	2.16	2.79	34	35
A. robustus	47.7	530	2.50	3.24	40	40
A. boisei	46.1	515	2.50	3.22	40	40
'Broad'						
H. habilis	40.5	631	3.37	4.30	54	53
Early						
H. erectus (*H. ergaster*)	58.6	826	3.33	4.40	53	55
H. sapiens	44	1250	6.28	8.07	100	100

lobes. The sulcal and gyral patterns of these resemble later *Homo*, with the parts including the language centres particularly enlarged. For example, the frontal lobe of 1470, OH 16 and, where preserved, OH 7 is very like later *H. erectus* with a prominent Broca area (Tobias, 1987). In addition, at least some specimens show the human pattern of right frontal–left occipital asymmetry that correlates with handedness and some aspects of hemispheric dominance. On the other hand, Holloway and Falk (1983) point to distinctive, australopithecine or 'pongid-like' features of the ER 1805 endocast, so providing further evidence for heterogeneity within early *Homo*.

ER 1805 aside, all this points to a relatively advanced (although not necessarily modern) central nervous system in early *Homo*, and Tobias (1991) argues that artefacts, language and brain evolution are intimately related in a reinforcing way. He sees tools as essential for *Homo*'s survival, and their production facilitated by socialisation and language. These developments promoted brain expansion and elaboration, so feeding back to further reinforce language development, socio-cultural transmission and further brain evolution. However, this interpretation presupposes an entirely human mode of behaviour and adaptation (see below). We should beware of attributing all artefacts exclusively to *Homo*, since there is no way of establishing the minimal neural basis for regularly patterned tool making of the kind recognised in the archaeological record. The balance of evidence points to cortical reorganisation in *Australopithecus* (see chapter 5) and to brain traits paralleling *Homo* in the robusts (Holloway, 1988). There accordingly appears no reason to exclude robust australopithecines from making at least some of the tools recognised at these sites.

There are other grounds for caution. The commingling of artefacts and bones is the main reason for thinking that the latter represent hominid food remains, and that tools are devices for procuring this. It has often been uncritically assumed that early hominids were proficient hunters, and many behaviourial reconstructions stress the hunting capacities of early *Homo*. However, such reconstructions ignore the realities of recent tropical hunter–gatherers, most of whose diet consists of plant items, the physical characteristics of early hominids (in many cases small and lightly built, and so physically weak) and the many factors that may have influenced site formation. Recognition of these, largely as a result of the work of Binford (1981), has led to a critical reassessment of early hominid behaviour. Binford argues that many so-called living floors accumulated over a period of time through a variety of causes (e.g. carnivore activity, water action) so that hominid activity plays, at best, a minor role, and certainly not one sufficient to justify claims of hunting or significant meat eating.

These arguments have prompted several others to reappraise the evidence; Isaac and Crader (1981) provide a careful review of the relevant data from early African sites. They conclude that at most sites the deposition of artefacts and bones was essentially isochronous (i.e. over a short period) with hominid involvement highly probable, and that the sites therefore provide evidence of meat eating. This is further supported by the intimate association of bone fragments with tools, and of undoubted cut marks on bones at Koobi Fora and Olduvai. However, the extent, complexity and purpose of such activity are disputed and could range from opportunistic scavenging and marrow extraction from the remains of carnivore prey to food sharing that involved purposeful hunting, social cooperation and/or pair bonding. Clear evidence for systematic hunting is lacking at the very earliest sites (Isaac and Crader, 1981), and the pattern of Koobi Fora remains is more consistent with scavenging (Shipman, 1983); most cut marks are *over* those produced by carnivore teeth, indicating later hominid access to the carcass.

Attempts to estimate directly the relative importance of meat and plant foodstuffs in early hominid diets (Boaz and Hemple, 1978; Sillan and Kavanagh, 1982) have proved inconclusive, but it is likely that plant items were the dominant proportions. Although there have been claims for the early occurrence of fire *c.* 1.3 my ago (Gowlett *et al.*, 1981), the evidence is inconclusive, and meat was probably eaten raw. Much flesh is extremely tough, and availability would be limited to those parts which could be masticated by hominid teeth and jaws, a relatively small proportion. Walker (1981) has suggested that nutritional mistakes may well have occurred: KNM ER 1808 (Koobi Fora) is a partial skeleton with pathology compatible with excess vitamin A, perhaps resulting from eating quantities of liver which may have been attractive because of its softness and the relative ease with which it could be chewed.

The Oldowan assemblages reveal no evidence of imposed 'style', and tools were determined principally by the nature of the raw material, functional requirements and a least-effort sequence of operations; compared with later tools they are undoubtedly simple and rudimentary. Their forms are therefore best viewed as 'one-off' responses to immediate needs rather than imposed into discontinuous stylistic types by their maker: there is no evidence of 'design'. Wynn and McGrew (1989) have argued that the minimum cognitive skills of proximity, boundary and order required for Oldowan production are no greater than those displayed by modern apes and that the tools accordingly reflect an ape, not human, adaptive grade. Nonetheless they reveal behaviours not matched by any non-human primate: while chimps use cobbles to crack nuts, no ape regularly flakes stones in the wild, although an orang has been taught to do so in captivity (Wright, 1972): regular flake removal requires practical knowledge of the stones' fracturing properties and the direction and force of blow required. Moreover, there is evidence of selection of raw material: some Olduvai tools are made on pebbles foreign to the locality, and at least some appear to have been brought from several kilometres away (Toth, 1985).

In addition, meat eating, although less than some accounts suggest, was undoubtedly greater than in any modern ape. Even if scavenged this implies success in competition with carnivores. Wynn and McGrew (1989) suggest deception strategies, now well documented in mating rivalries, etc. (Byrne and Whiten, 1988) as important here, and a factor in the evolution of more human intelligence. Perhaps the best categorisation of Oldowan intelligence is Wynn and McGrew's (1989) description of an ape-grade adaptation, but one pushed to its very limits. Other aspects of early hominid mental skills are considered in chapter 7.

There have been many attempts, all speculative, to reconstruct the social behaviour of early hominids, often flawed by an unrealistic reliance upon modern human parallels with marked sexual division of labour, and an emphasis on male sociality resulting from cooperative hunting which, as noted above, is probably exaggerated.

Alternative models are those of Isaac (1978) and Lovejoy (1981) who stress the varied, uncertain nature of food resources in seasonal savannah environments. Isaac emphasises the survival value of food sharing in such conditions, leading to provisioning, home bases and the monogamous family social unit, as well as tools to make containers for food transport. Lovejoy (1981) posits an earlier (pre-Pleistocene) evolution of monogamy and the nuclear family following the effects of Miocene ecological change on hominoid species with slow maturation, longer infant dependency, longer birth intervals and so fewer offspring compared with cercopithecoids.

Homin(o)ids would need to counter this reproductive handicap by increased parental investment and other strategies to ensure greater offspring survival and/or reduced birth spacing. A basal hominoid behavioural adaptation involved the evolution of separate foraging ranges for the sexes to avoid males competing with females and infants for food resources. Monogamous pair bonding would be favoured in such a situation, since bonding (and reproductive success) would be reinforced by males provisioning females and infants, leading to enhanced offspring survival, allowing a reduction in birth spacing compared with behaviours where the mother exclusively feeds herself and the infant. Lovejoy argues that bipedalism evolved early as a locomotor pattern that allowed more effective food transport and so aided provisioning.

There are several difficulties with this model. It posits modern human social structure (for which there is no evidence) at a time when all available evidence is manifestly non-human. Monogamous pair bonding correlates with reduced sexual dimorphism, since bigger bodied males gain no reproductive advantage over smaller ones and so the two sexes converge in size. Yet all available fossil evidence suggests early hominids were characterised by pronounced sexual dimorphism. Most of all, the model is driven by the assumption that early hominids were human-like in their long growth and development periods and helplessness as infants. Again, evidence indicates the opposite was true, with rapid growth in *Australopithecus* for example.

Foley and Lee (1989) have attempted to embed reconstruction of early hominid social structure in the context of primate socio-ecology, by considering the spectrum of behaviour and its morphological and ecological correlates in extant primate species. They see basal hominids as open-habitat forms with large mixed-sex groups based on male kinship, and females bonding with individual males or all males. This interpretation is compatible with the two- species model at Hadar, but not with one species, From the fossil evidence this would have to be markedly dimorphic, which implies intense reproductive competition between males and social groups more like the orang or gorilla, but it is difficult to reconcile these theories, which require the presence of abundant foods, with the palaeo-environmental evidence.

Foley and Lee (1989) posit a socio-ecological divergence of later Pliocene hominids with one trend—that of the robust australopithecines—leading to the exploitation of widely distributed high-volume/low-quality items. These allow for large groups loosely linked by male kin bonds, important for defence against predators and conspecific intrusions, and within the group unrelated females associating with individual males, i.e. a harem structure.

Foley and Lee (1989) see early *Homo* representing a shift towards increased exploitation of higher quality, larger sized although unpredictably distributed food, leading to increased male cooperation, both to obtain the food and to defend females from intruder males. In addition, male–female bonding also developed, conferring advantages of provisioning and reduced mortality from predation or intraspecific conflicts.

Such models help illuminate the fossil record but are necessarily speculative, and as such depend heavily upon knowledge of early hominid morphology, environments and phylogeny as significant inputs. As such, they are likely to change appreciably over the next few years.

CHAPTER SEVEN

LOWER AND MIDDLE PLEISTOCENE HOMINIDS

7.1 Introduction

The Lower and Middle Pleistocene (*c.* 1.5–0.2 mya), spanning the greater part of the Quaternary, document a phase of human evolution that contrasts in several respects with the preceding Pliocene and basal Pleistocene. Hominid remains are more frequent but morphologically less variable, and contrasts are particularly marked with the expanded diversity around the Plio-Pleistocene boundary. The implication is of fewer hominid species and, indeed, early *Homo* specimens (*H. habilis* and *H. rudolfensis*) disappear from the record after about 1.5 mya and robust australopithecines follow (see chapter 6). The reasons for this are obscure, but it appears to correspond generally with a phase of high faunal turnover: numerous Pliocene relicts become extinct about this time and the remaining fauna takes on a more modern aspect. Such gross faunal changes probably reflect major ecological shifts resulting from Pleistocene climatic fluctuations (see below).

The upshot is a single hominid morphology, that of *Homo erectus* grade, as the dominant fossil form over the greater part of this period. The material is not uniform, in fact it shows considerable variation, but there is no good evidence, as in the preceding phase, of coexisting, contrasting forms implying sympatric hominid species. Some of the observed variation appears to be temporally directed (see below for contrasting interpretations of this) while other variation is probably due to individual diversity. A further component almost certainly represents polytypism or regional differentiation. In the earlier Lower Pleistocene fossil remains are confined to Africa but after about 1.0 mya they begin to appear elsewhere, initially within the tropics, later in higher latitudes. As hominids expanded their range, marginal populations were likely to have come under more intense local selection pressures, resulting in parochial features due to local adaptations and perhaps periodically interrupted gene flow from more

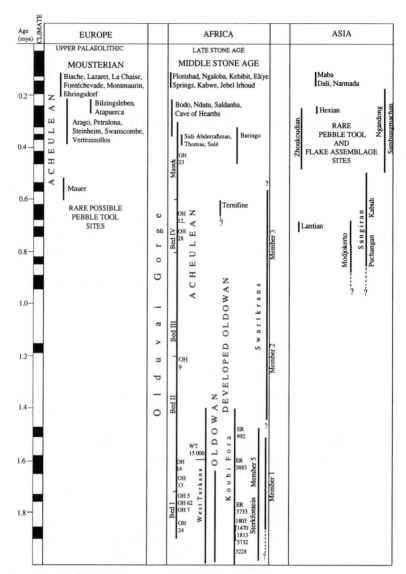

Figure 7.1 Final Pliocene, Lower and Middle Pleistocene time-scale. Probable time ranges of some important hominid sites and artefact assemblages are indicated. Climatic oscillations revealed by oxygen isotope analysis (deep sea core record) at left. Dark blocks represent periods of increased ice sheet formation and reduced ocean volume (lower temperatures), light blocks represent reduced ice sheets and expanded ocean volume (warmer conditions). Changes prior to 1 mya are less well defined; the magnitude of climate contrasts increased appreciably after *c.* 0.8 mya (Middle and Upper Pleistocene).

central populations. Such components of variability need to be kept in mind when evaluating the fossil record.

The Pleistocene was a period of repeated climatic change, with effects particularly pronounced in high latitudes. Contrasting temperatures produced prolonged cold, sometimes glacial, conditions over much of Europe and Asia with dramatic effects on both flora and fauna. With so much water locked up in expanded ice sheets, sea levels fell, linking Britain to Europe and, more significantly, exposing the Sunda shelf and linking Indonesia to the Asiatic mainland. Outside glacial and periglacial regions rainfall decreased, leading to phases of aridity within parts of the tropics. Such colder 'glacial' periods were separated by interglacials, usually warmer than at present, and with higher rainfall within the tropics. Early

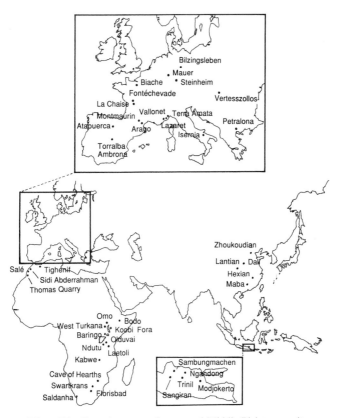

Figure 7.2   Some important Lower and Middle Pleistocene sites.

climatic reconstructions, based largely upon European data, suggested a four- (later five-) fold glacial sequence, separated by warmer interglacials. However, the terrestrial record is an erosional one, with each glacial phase largely removing evidence of earlier fluctuations. Palaeo-temperature reconstructions based on depositional deep sea core data provide a much more complete record and show the earlier framework to be hopelessly inadequate: there appear to have been at least a dozen glacial phases over the last 700 000 years alone. Such multiple climatic oscillations undoubtedly had a major effect upon hominid populations, as on all other species.

Dating is a problem for much of the period covered by this chapter. It is now usual to take the Brunhes/Matuyama palaeo-reversal (c. 0.72 mya) as a convenient if arbitrary interface between the Lower and Middle

Table 7.1　Some Lower and early Middle Pleistocene hominid sites.

Site	Age (mya)	Fossils	Species
Africa			
Koobi Fora, Kenya (East Turkana)	1.6–1.8	Crania (3733, 3883), mandibles (820, 992), postcranials, skeleton (1808)	*H. ergaster*/*H. erectus*
Nariokotome, Kenya	1.6	Skeleton (WT 15000)	*H. erectus*
Other West Turkana sites	1.5–1.8	Crania, mandibles	*A. boisei*
Olduvai Gorge, Tanzania			
Upper Bed II	1.2	Cranium (OH 9)	*H. erectus*
Beds III–IV	0.7–1.0	Cranium (OH 12), hip bone and femur (OH 28), mandibles (OH 22, 23, 57)	*H. erectus*
Swartkrans, South Africa			
member 2	1.0–1.5	Mandible (SK 15)	*H. erectus*
Tighenif, Algeria (Ternifine)	0.6	Mandibles (TI–III)	*H. erectus*
Asia			
Trinil, Java	0.5–?0.7	Skull cap, femur	*H. erectus*
Sangiran, Java			
Kabuh	0.5–0.7	Many crania, mandible fragments	*H. erectus*
Puchangan	0.7–?1.0	Few finds	*H. erectus*
Modjokerto (Perning), Java	0.7–?1.0	Cranium	*H. erectus*
Lantian, China			
Gongwangling	*c.* 0.7	Cranium	*H. erectus*

Pleistocene, and at the outset of the period the relatively well-dated East African sequence provides a chronological baseline. However, major problems occur elsewhere: many fossils are from localities where absolute dating techniques are not applicable or, if dates are available, they are subject to considerable uncertainty or error. By the later Middle Pleistocene the time range is generally too young for accurate potassium–argon dating, and other methods have yet to provide a detailed chronology. Faunal and geological comparisons, once confidently used to establish at least a relative sequence within the old fourfold glacial framework, can no longer be relied upon because the recognition of multiple climatic oscillations means that a warm period identified at one locality may not correspond to those recognised elsewhere, and so even relative chronologies may be suspect, particularly if they attempt to span widely separated areas. These problems and uncertainties need to be borne in mind when assessing the temporal and evolutionary relationships of many of the specimens described below.

7.2 Early *Homo erectus*

The earliest specimens generally regarded as *H. erectus* are from East Africa, especially around Lake Turkana. On the eastern side of the lake two crania, KNM ER 3733 and KNM ER 3883 from the KBS and Okote members dated to *c.* 1.78 and 1.59 mya respectively, are usually assigned to this species (Walker and Leakey, 1985) or to the recently identified *H. ergaster* (Wood, 1991) which is considered to be of an equivalent grade. Compared with other early hominids these specimens are large brained (*c.* 850 and 800 cm^3 respectively) and exhibit characteristic *H. erectus* neurocranial proportions: the cranial base is wide and the vault walls relatively vertical in their lower portions so that from the rear the braincase appears arched rather than bell shaped. The occipital is large and sharply angled, with a well-marked area for the neck muscles bounded by a distinct nuchal ridge. The temporal lines are also distinct and slightly raised, especially anteriorly, but because of the skull's breadth are always well separated, so that there is no trace of a crest. The prominent brow ridge has a semicircular outline above each orbit; it is thicker in ER 3883. There is also some midline thickening (keeling) of the vault in the frontal and parietal regions. In ER 3733 the vertically orientated frontal is set off from the brow ridge by a distinct groove (sulcus); in ER 3883 the torus shelves directly into the more retreating frontal bone.

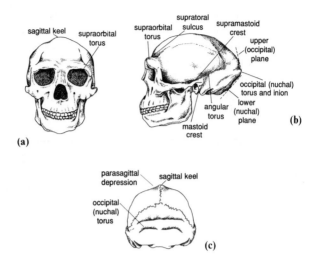

Figure 7.3   Reconstruction of Zhoukoudian cranium to show main *H. erectus* features.

KNM ER 3883 consists only of a braincase, but the 3733 cranium is intact, so providing information about the face and upper jaw of early *H. erectus*. Compared with other early hominids, the face is relatively short and flat, broader in its upper portion but with light zygomatic arches. The palate is small and the cheek teeth reduced, given the size of the cranium. Other fossils from east Turkana, mainly mandibles and postcranial bones, may also represent *H. erectus* but are either too incomplete or lack characteristic features for definite assignment.

This is not the case with the Nariokotome, west Turkana, specimen KNM WT 15 000, a virtually complete skeleton of an *H. erectus* youth, about 12 years at death, and dated to *c.* 1.6 mya. The individual was tall and ruggedly built with a powerful physique but with largely modern postcranial proportions. In contrast to other early hominids, including the more or less contemporary OH 12 and ER 3735, the relative lengths of the arm and leg (measured by the intermembral index) are virtually identical with the modern human average. Height and weight have been estimated at *c.* 1.66 m and *c.* 55 kg respectively (Leakey and Walker, 1989). The individual is considered male on pelvic proportions and facial size, and is bigger and stronger, despite youth, than the adult 3733 specimen. Cranial capacity is *c.* 900 cm^3.

While overall body proportions and many individual features are surprisingly modern, there are some interesting and significant contrasts of

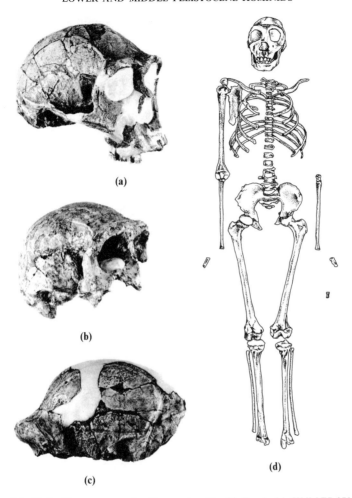

Figure 7.4 Early *H. erectus* grade. *H. ergaster* (Koobi Fora): (a) KNM ER 3733; (b) KNM ER 3883. *H. erectus*: (c) OH9: (d) KNM WT 15000 (from a photograph in Day, 1986).

the trunk and thigh (Leakey and Walker, 1989). The spinous processes of the upper vertebrae (neck and upper chest) are nearly horizontal, as in apes, rather than sloping downward as in humans, so increasing the power arm of the muscles attaching there. Similarly, the rib case is conical or bell-shaped, rather than barrel-like as in modern humans. The specimen resembles Lucy in these two interrelated traits, suggesting that the upper trunk lagged behind other parts of the body in evolving full adaptation to truncally erect bipedalism.

The hip bone shows typical *erectus* features (see below), but these are also seen in early, possibly non-*erectus* forms such as ER 3228 (see chapter 6), and the sacrum and complete pelvis are transversely narrow. The femur head is large, but there are australopithecine-like traits in the very long femoral neck and its low angle with the shaft. As with the hip bone, a femur (KNM ER 1481) from the Upper Burgi member (*c*.1.9 mya) closely resembles that of the Nariokotome youth, and may represent earlier *H. erectus* or a cranially distinct but postcranially similar form (*H. rudolfensis*).

Also ruggedly constructed is the OH 9 cranium from upper Bed II of the Olduvai Gorge and younger than the above specimens at *c*.1.2 mya (Leakey, 1961a; Rightmire, 1979b; 1990). The braincase at *c*.1000 cm^3 is larger than the Koobi Fora crania and more heavily constructed, with thick vault bones, strong temporal muscle markings and a massive supraorbital torus. Possible evidence of early *H. erectus* in southern Africa is provided by the SK 847 craniofacial specimen from a pocket of breccia within member 1 at Swartkrans. Both identity and age are uncertain (the specimen is also widely regarded as *H. habilis* or early *Homo* sp., see chapter 6) but probably falls within the 1.5–2.0 mya range.

The above constitute the main specimens of early *H. erectus* from sub-Saharan Africa. There are also a number of other fossils which are usually regarded as later members of the some species. They include: two mandibles (BK 67 and BK 8518) from the Kapthurin beds near Lake Baringo in Kenya (Leakey *et al.*, 1969; Van Noten, 1983; Wood and Van Noten, 1986); and from Bed IV of the Olduvai Gorge (0.7?–1.0 mya) a partial, relatively thin cranial vault (OH 12) (Rightmire, 1979) and a hip bone and femoral shaft (OH 28) (Day, 1971). The hip bone is large and ruggedly constructed, so matching that of the earlier Nariakatome skeleton, and the femur shows some detailed similarities with other *H. erectus* femora from Asia (see below). Also from Olduvai are several mandibles (Rightmire, 1990): OH 22 is a nearly complete right mandible body with P_3–M_2 from side gorge deposits equivalent to Bed III or IV: OH 23 from the Mazek beds (0.4–0.6 mya) is a fragment of left corpus with P_4 (broken)–M_2, while OH 57 from Beds III–IV is a left corpus fragment with P_4 and M_1.

A further Swartkrans fossil is a relatively small, lightly built mandible (SK 15) originally called 'Telanthropus' and from another, later breccia pocket with a younger fauna (perhaps around 1.0–1.5 mya) than the SK 847 composite specimen. The jaw dimensions are within the *H. erectus* range, as are the dimensions of the two preserved molar teeth.

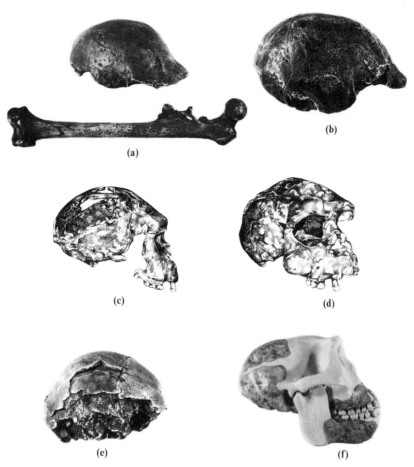

Figure 7.5 SE Asian (Java) *H. erectus*. (a) Trinil skull cap and femur; (b) Sangiran I cranial vault; (c) and (d) two views of Sangiran 17. (e) Modjokerto infant cranium; (f) reconstruction of early Javanese *H. erectus* based on Sangiran 4 rear vault and palate, and B mandible.

7.2.1 *Asia*

Overlapping with at least some of these later African fossils are numerous specimens from Asia, so providing evidence of hominid expansion beyond Africa. The first discoveries of *H. erectus* were, in fact, from Asia, and the majority of specimens are still from that continent. There are two primary foci: Indonesia (Java) and China, especially close to Beijing.

7.2.1.1   *Indonesia*.   The original discovery of *H. erectus* was made in 1891 in Trinil, a site on the Solo River in central Java. Dubois recovered the upper part of a cranial vault (calotte), a femur and isolated teeth (subsequently shown to be from an orang). The specimens were found on a sandbank over a distance of 400 m or so, and therefore not in close association. Sufficient of the skull cap is present to show a strong brow ridge, narrow retreating frontal, flat vault and angled occipital with strong nuchal markings. Dubois was struck by the disparity between the apparently primitive skull cap and modern femur, hence the name *Pithecanthropus erectus*, erect apeman; the species is now included in *Homo* as *H. erectus*.

The femur shows a pathological outgrowth of bone (exostasis) in response to injury, but is otherwise modern in appearance and lacks some of the detailed features of other *H. erectus* femora. While chemical analysis confirms the femur's antiquity, it may well be rather younger than the skull cap, and there is nothing to directly associate the two specimens. The site of discovery, on a river bend, is one where numerous objects would be redeposited by the river current, and both specimens are abraded. Despite extensive excavation, no further hominid material has been found at Trinil. It is therefore ironic that Dubois early championed bipedalism (as opposed to brain expansion) as the primary hominid adaptation on the basis of possibly incorrect evidence.

The Sangiran area, some 40 miles west, has produced many fossil specimens. An eroded volcanic dome has left a series of beds, the oldest towards the centre of dome. Several series of deposits are exposed, including the Middle Pleistocene Kabuh beds, also exposed at Dubois's original site of Trinil, with a prominent marker bed of chalky sands and gravels, the Grenzbank, towards the base. Below the Kabuh series are the Puchangan or Djetis beds, dating from the early Middle or possibly late Lower Pleistocene, and over the Kabuh beds are the late Middle/Upper Pleistocene Notopuro beds (Pope, 1988).

Over the last 50 years or so numerous hominid specimens have been recovered from the Sangiran area; early discoveries were made by von Koenigswald, while Indonesian workers, especially Sartono and Jacob, have reported more recently recovered specimens, some not yet fully described. Virtually all have been found as a result of retainers paid to local collectors, rather than through controlled excavation. There is therefore a lack of detailed stratigraphic and contextual data for the specimens when compared with, for example, the African fossils. Circumstances of recovery also explain why virtually all specimens are easily recognised crania, with little in the way of postcranial evidence.

Most specimens whose provenance is known are from the mid to upper part of the Kabuh (Trinil) beds, well above the Grenzbank and therefore probably between 0.5 and 0.75 my in age. A few specimens are from towards the base of the Kabuh formation, and one or two, including the massively built Sangiran 4 cranium and some mandibular remains, appear to be from the underlying Puchangan beds, for which an age of c. 0.8–1.0 my appears not unreasonable. Probably of similar age is the infant cranium recovered by von Koenigswald from Modjokerto (Perning), further to the east nearer Surabaya. This specimen provides important information on the pattern of cranial development and shows the early appearance of characteristic H. erectus features, narrow and retreating frontal, incipient brow ridge and angulated occipital with strong developing nuchal torus.

Close to Sangiran is Sambungmachan, where in 1973 Jacob recovered a cranium of broadly H. erectus appearance but with higher more inflated frontal and expanded occipital region compared with the Trinil and Sangiran remains (Jacob, 1973; 1976). Dating is uncertain, with some evidence pointing to a Middle Pleistocene (Kabuh) age, and other evidence indicating an Upper Pleistocene date. The specimen shows undoubted similarities to the much larger sample recovered in the early 1930s from Ngandong downriver from Trinil. This site yielded 12 crania all lacking face and jaws and with the base broken, suggesting artificial damage, together with two tibiae. The crania are larger than those of Sangiran H. erectus, with cranial capacity of c. 1000–1250 cm^3 reflected in higher vaults and more vertical frontals and with the sides laterally expanded (Weidenreich, 1951). The specimens were recovered from the Notopuro deposits of Upper Pleistocene age, and are usually regarded as either advanced erectus (H. erectus soloensis) or a primitive subspecies of Homo sapiens (H. sapiens soloensis) (see below).

The Javan fossil sample is an extensive (and growing) one which should yield valuable information on Asian H. erectus. Unfortunately its value is limited by the lack of contextual information, and data about some of the more recently recovered specimens are particularly sparse. Brain size ranges from 775 cm^3 (the first Sangiran discovery) to around 1080 cm^3 for typical H. erectus and up to 1255 cm^3 for the Ngandong crania. The vault bones are thick and there is often a frontal and/or parietal keel along the middle of the vault, either as a continuous ridge or as a series of knobs and bumps; the sides of the vault are reinforced by a thickening at the rear of the parietal and temporal bones termed the angular torus, and the occipital bone is expanded and sharply angulated with an extensive nuchal area separated from the upper portion by a prominent ridge or torus. The brow ridge is pronounced and shelves into the retreating frontal bone, which is

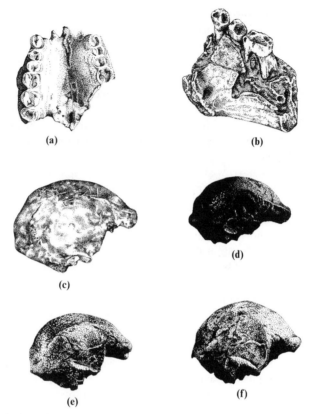

Figure 7.6   SE Asian *H. erectus*. (a) Sangiran 4 palate; (b) Sangiran 6 mandible (Megan-thropus II); (c) Sambungmachan cranium; (d–f) Ngandang (Solo) crania: (d) Ng. 6; (e) Ng. 1; (f) Ng. 12.

also narrow, so that there is marked constriction behind the orbits. The specimens span a considerable period of time (probably several hundred thousand years), although not as long as some have suggested: there is no good reason to suppose that any of the specimens is more than 1 my old at most. Several workers have detected heterogeneity within the material; Weidenreich and von Koenigswald were impressed by what they considered to be more rugged and primitive features in the few specimens from the Puchangan beds and identified two hominid species within that sample. One was an early, rugged *H. erectus* form, *Pithecanthropus robustus* (Weidenreich, 1945) or *Pithecanthropus* (later *Homo*) *modjokertensis* (von Koenigswald, 1950; 1975).

Weidenreich (1954) also considered some especially large mandible remains from Sangiran to represent a larger and more primitive hominid he called *Meganthropus palaeojavanicus* and suggested an evolutionary sequence from *Meganthropus* through *P. robustus* to *P. erectus*. Others have considered the *Meganthropus* material to provide evidence of another hominid in Java: Robinson (1953) argued that it represented a robust australopithecine, whilst Tobias and von Koenigswald (1964) drew attention to similarities between the Olduvai Bed I *H. habilis* material (OH 7 mandible) and *Meganthropus* and the Olduvai Bed II (OH 13) specimen and the Sangiran 4 maxilla and B mandible, suggesting that the Puchangan material possibly represents Javan representatives of *H. habilis* or a similar species.

All these claims for additional species in Java appear wide of the mark. Although the *Meganthropus* mandibles are large, their cheek teeth are too small to be robust australopithecine, and the undoubted resemblances between the Olduvai and Javanese specimens in teeth and jaws do not demonstrate the existence of *H. habilis* in Java. Both species of early *Homo* overlap in jaw characters and dental dimensions, so that it is difficult if not impossible to allocate isolated fragmentary specimens, whether from East Africa or Asia, to one or other species with any certainty. However, these similarities do not extend to the cranial vault, and there is no Indonesian evidence for the characteristically thin-vaulted, more globular braincase of *H. habilis* or early *Homo* sp; the Javanese neurocrania are ruggedly constructed and typically *erectus* in their proportions.

Neither does there appear to be any compelling reason for distinguishing between closely related *H. erectus*-like species between the Puchangan and Kabuh beds. Given the small number of specimens definitely known from the former, the few contrasts between the two samples could easily reflect individual variability. The original distinction of Weidenreich, for example, based primarily on robustness appears to have been eroded with the recovery of more material; Sangiran 17, from well up in the Kabuh beds, is as robust and strongly constructed as some of the earlier specimens.

In fact, the primary impression of the Javanese sample is one of similarity and the repeated occurrence of particular features. This is not surprising, given its geographical context. Indonesian *H. erectus* communities were on the extreme periphery of the hominid range and as such were subject to strong local selection pressures. Moreover the Indonesian archipelago would only periodically be accessible from the Asian mainland, during periods of low sea-level (glacials in higher latitudes) when the Sunda shelf would be exposed. During warmer, interglacial periods, sea levels rose, isolating the islands and the hominid groups. These are just the circum-stances in which founder effect, or a restricted gene pool, combined with local selection pressures would be likely to have a stabilising effect and lead to the persistence of distinctive features in the Javan material. The persistence of such 'local' features and the context of the Javanese fossils have direct relevance for the broader issues of whether *H. erectus* is essentially an Asian species, and whether it represents a period of stasis in human evolution (see below).

While it may be difficult to distinguish between the Puchangan and Kabuh fossils, there is a stronger case for separating these forms from the Ngandong and Sangbungmachan specimens. Even here, given the limited nature of the material, contrasts are limited: an increase in cranial capacity and reproportioning of the vault, a reduction in the skull reinforcement system (sagittal keeling, angular torus) and a smaller nuchal area and diminished nuchal ridge. Most debate about the Ngandong crania has been not morphological or phyletic but taxonomic—whether they are more appropriately considered as an advanced form of *H. erectus* or primitive *H. sapiens* (see below). Continuity with earlier *H. erectus* is provided by the same distinctive cranial features (sagittal keeling, nuchal characteristics), albeit in an attenuated form (Weidenreich, 1951; Santá Luca, 1980).

7.2.1.2 *China.* Evidence for adjacent and probably parent *H. erectus* populations on the Asian mainland is provided by fossil remains from China. The earliest is probably the Gongwangling cranium from Lantian, central China, a much battered upper vault with fragments of the face. The vault is very like some of the earlier Javanese specimens, and faunal remains from Lantian are similar to those of the Djetis fauna remains from the Puchangan beds on Java. The date of *c.* 0.7 mya for the site has implications for the date of the Javanese *erectus* specimens, for it suggests that even the earliest Indonesian fossils are appreciably later than is often thought.

By far the largest and best-known Chinese hominid sample is that from Zhoukoudian (ZKD) near Beijing. Here a deep cleft in a large vertical cliff has been systematically excavated since the 1920s and has yielded much faunal and archaeological material, as well as hominid remains. The bulk of the specimens, recovered before 1940, were lost in World War II but are well known through casts and a series of masterly monographs by Weidenreich. Most of the material is cranial, mainly braincases, but there are some face and jaw fragments and many teeth, as well as some postcranial bones.

The crania are broadly similar to the Javanese specimens with typi-cal *erectus* features of expanded occipitals, extensive nuchal area and pro-

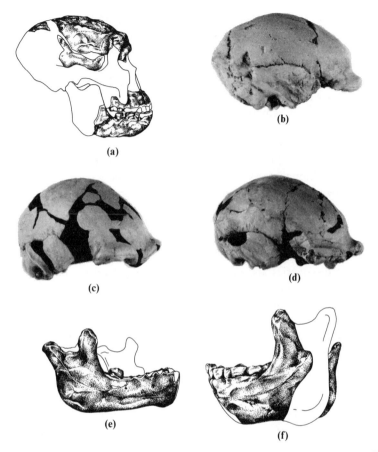

Figure 7.7 Chinese *H. erectus*. (a) Reconstruction based on the Lantian (Gongwangling) skull cap and maxilla, and the Lantian (Chenchiawo) mandible. (b–f) Zhoukoundian specimens: (b) ZKD skull III; (c) ZKD skull X (Locus LI); (d) ZKD skull XI (Locus LII) (Locus E); (e) mandible H1 (female); (f) mandible G-1 (male) (see also Figure 7.3).

nounced torus, marked brow ridges, distinct mid-sagittal keeling and an angular torus. However, there are also contrasts: cranial capacity is greater (915–1225 cm³) and is reflected in higher vaults with more inflated, wider frontals and larger parietal bones, and the crania are generally less massive than the biggest Javanese specimens (e.g. S17).

Mandibular remains are similar in size to, or smaller than, those from Java, and no palatal fragment is as large as the S4 specimen. Dentally the ZKD samples show a high incidence of incisal shovelling (a trait characterising some modern Asian populations) and the cheek teeth (premolars and molars) are reduced compared with many other early hominids so that they just fall within the range of large-toothed contemporary populations such as the Eskimos or Australian aborigines.

Postcranially the ZKD specimens show resemblances to the African material described above. The hip bone fragments are robustly built and the femoral and tibial remains show detailed morphological features that match across the two continents. The Chinese specimens are, however, short, indicating a total height of only around 5 ft or so, contrasting with the African specimens. Such contrasts may reflect polytypic variation or, possibly, nutritional stress in high latitudes (see below).

The growing number of *H. erectus* specimens from East Africa around 1.8–1.5 mya suggests the early and probably rapid evolution of the species within that explosive burst of hominid diversity during the later Pliocene and basal Pleistocene. There have been claims that *H. erectus* originates through phyletic (anagenetic) evolution within a single hominid lineage, but recent discoveries and dating developments effectively demolish this view. The Turkana sequence shows *H. erectus* present by 1.78 mya (ER 3733) but most specimens are rather younger (1.6–1.7 mya). Most *H. rudolfensis* specimens are only slightly older (1.85–1.9 mya) and one (ER 3891) is contemporaneous (1.77 mya). *H. habilis* at Koobi Fora is also dated between 1.85 and 1.9 mya but at Olduvai extends down to around 1.5 mya in Bed II.

The distribution, type and scale of the morphological contrasts point to *H. erectus* as a distinct clade resulting from a speciation event in the later Pliocene, and with other early *Homo* species persisting for some time thereafter. Neither *H. habilis* nor *H. rudolfensis* provides an appropriate ancestor for *H. erectus*. *Habilis* displays a similar facial structure, masticatory apparatus and dentition, but is associated with a small, lightly constructed vault and remarkably primitive postcrania. *H. rudolfensis* shows an expanded cranial vault, but of different form, and face, jaws and dentition are decidedly non-*erectus*. Postcranial remains are similar and, if correctly identified as *H. rudolfensis*, point to a period of recent shared ancestry, perhaps around 2 mya. If not correctly associated, these postcrania (e.g. the 3228 hip bone) push *H. erectus* back to at least 2 mya and a correspondingly earlier speciation origin. The clear implication is of multiple coexisting hominid species (quite apart from *A. robustus/boisei*) during the later Pliocene, presumably reflecting niche differentiation although details are obscure. In any event, the morphology—large,

Table 7.2 Autapomorphies of *H. erectus*.

Wood (1984)	Andrews (1984)
1. Occipital torus with sulcus above	1. Frontal keel
2. Angular torus and mastoid crest	2. Parietal keel
3. Sulcus on frontal behind torus	3. Angular torus on pariental
4. Proportions and shape of occipital bone	4. Inion well separated from endinion (nuchal markings)
5. Relatively large occipital arc	5. Thick vault bones
	6. Mastoid fissure
	7. Recess between entoglenoid and tympanic plate on cranial base

strongly constructed vault combined with relatively small face, teeth and jaws, and strongly derived postcrania of broadly modern form—sets the *H. erectus* clade apart from other hominids.

In recent years workers have increasingly questioned whether the early African material represents *H. erectus* as it is known from Asia. Andrews (1984) and Wood (1984), for example, argue on cladistic grounds that *H. erectus* is an exclusively Asiatic taxon and that the material there possesses autapomorphies not present in the African material and also not present in *H. sapiens*. They conclude that the African specimens are more clearly implicated in the ancestry of *H. sapiens* and that Asian *H. erectus* is not. Table 7.2 lists the autapomorphic features of *H. erectus* cited by Wood and Andrews.

There are several problems with this analysis. The first is that some of the claimed Asian autapomorphies are, in fact, present in some of the early African specimens. For example, KNM ER 3733 possesses a supratoral sulcus (Wood, 1984) and a frontal keel (Andrews, 1984). Moreover, these features are not invariably found in Asian *H. erectus*; the Trinil-type specimen, for example, lacks a post-toral sulcus and frontal keel. The best that can be demonstrated is that each Asian specimen possesses at least one of the characters listed, and that the group as a whole shows in combination the features listed by Wood (1984) and Andrews (1984). In other words, *H. erectus* is polythetic: it is not necessary for every individual specimen to possess all listed characters.

Given the polythetic nature of Asian *H. erectus* and the presence of at least some autapomorphic features in the African material, there is no reason to exclude the latter group from *H. erectus*. To do so is simply to assert that *H. erectus* is an Asian species, not because it is morphologically distinct, but because the bulk of specimens derive from Asia, which is arguing on geography, not morphology. The list of supposed autapo-morphic characters, and consequent uniqueness of the Asian sample, is further eroded by recognising that several of the listed features are not truly independent characters, e.g. frontal and parietal keels and an angular torus are all part of the skull reinforcement system and therefore best regarded as a single functional unit.

Wood (1991) has expanded and formalised the case for separating the Koobi Fora sample (ER 3733, 3883, 730, 820, 992) from *H. erectus* as *H. ergaster* (see chapter 6). In addition to the above list he cites primitive traits of the mandible and dentition (crown and root morphology) together with features of the vault, face, mandible and dentition that are shared with *H. sapiens*. Groves (1989) also argues that 3733 and 3883 are not *H. erectus*. However, Rightmire (1990), Leakey and Walker (1989), Turner and Chamberlain (1989) and Tobias (1991) also prefer to include the Turkana fossils within *H. erectus*. All agree that they differ from Asian (and later African) specimens in several ways: the issue is whether those differences are sufficiently consistent and fundamental to outweigh the many similarities and so warrant taxonomic recognition at the species level.

7.2.2 Stability or change in H. erectus

Fossil evidence increases in quantity and range during the later Middle Pleistocene (0.2–0.5 mya). In addition to Ngangdong and Sangbung-machen already noted above, there are a number of Asian finds, but the expansion is particularly marked in Europe and Africa north of the Sahara. Table 7.3 lists the more important specimens; some of these are fragmen-tary, but the more complete show numerous similarities with the classic samples from Java and Beijing. Facial proportions are still projecting and strongly constructed, with a prominent supraorbital torus, stoutly constructed mandible and dental proportions similar to those described above.

The recovery of the Koobi Fora and other early East African *H. erectus* specimens has encouraged the view of *H. erectus* as a relatively homo-geneous taxon representing a period of stasis in human evolution of the kind expected on a punctuational model (Gould and Eldridge, 1977; Stanley, 1979; 1981; Eldridge and Tattersall, 1982). Particular support for this view was provided by Rightmire (1981), who attempted unsuccessfully to identify trends of diagnostic or biomechanical significance in *H. erectus*. Rightmire's characters can be criticised, as can his sample composition—both early and late *erectus* groups include specimens usually assigned

Table 7.3 Some later Middle Pleistocene hominid sites.

Site	Age (mya)	Fossils	Species
Africa			
Kabwe, Zambia	0.1–0.2	Cranium, postcranials	*H. sapiens*
Bodo, Ethiopia	0.2–0.4	Crania	*H. sapiens*
Baringo, Kenya	0.3–0.5	Mandibles	*H. erectus*
Eliye Springs, Kenya	?0.1–0.2	Cranium	*H. sapiens*
Laetoli, Tanzania (Ngaloba)	0.1–0.15	Cranium	*H. sapiens*
Ndutu, Tanzania	0.2–0.4	Cranium	*H. sapiens*
Saldanha, South Africa	0.2–0.4	Cranium	*H. sapiens*
Makapansgat, South Africa (Cave of Hearths)	?0.2–0.4	Mandible	*H. sapiens*
Florisbad, South Africa	?0.1–0.15	Cranium	*H. sapiens*
Sidi Abderrahman, Morocco	*c.* 0.4	Mandibles	*H. erectus*
Thomas Quarry, Morocco	*c.* 0.4	Mandible, cranial fragments	*H. erectus*
Salé, Morocco	*c.* 0.4	Cranium	?*H. sapiens*
Kebibit, Morocco	0.15–0.2	Cranium	*H. sapiens*
Jebel Irhoud, Morocco	?0.1–0.2	Skull, cranial vault, other fragments	*H. sapiens*
Asia			
Ngandang, Java	?0.1–0.5	Crania, postcranials	*H. erectus*
Sambungmachan, Java	?0.1–0.5	Cranium	*H. erectus*
Zhoukoudian, China	0.25–0.5	Crania, teeth, postcranials	*H. erectus*
Hexian, China	0.25–0.3	Cranium, mandible	*H. erectus*
Dali, China	0.2–0.25	Cranium	*H. sapiens*
Maba, China	0.1–0.2	Skull cap	*H. sapiens*
Narmada, India	?0.2	Cranium	?*H. erectus*
Europe			
Mauer, Germany (Heidelberg)	0.5	Mandible	*H. erectus*/ *H. sapiens*
Arago, France	0.2–0.4	Crania, mandibles, postcranials	*H. erectus*/ *H. sapiens*
Petralona, Greece	?0.2–0.4	Cranium	*H. sapiens*
Atapuerca, Spain	0.2–0.35	Skull fragments, mandibles, teeth, postcranials	*H. erectus*/ *H. sapiens*
Vertesszollos, Hungary	0.2–0.4	Occipital	*H. erectus*/ *H. sapiens*
Swanscombe, UK	0.2–0.4	Cranium	*H. sapiens*
Steinheim, Germany	0.2–0.4	Cranium	*H. sapiens*
Bilzingsleben, Germany	0.2–0.3	Cranium	*H. erectus*/ *H. sapiens*
Biache, France	0.12–0.2	Cranium	*H. sapiens*
Lazaret, France	0.12–0.2	Teeth, skull fragments	*H. sapiens*
Fontéchevade, France	0.12–0.2	Skull fragments	*H. sapiens*
Montmaurin, France	0.12–0.2	Mandible	*H. sapiens*
La Chaise, France	0.1–0.25	Teeth, cranial and postcranial fragments	*H. sapiens*

to other taxa. For example, the early group includes OH 13, usually considered *H. habilis*, while the later sample incorporates Salé, Petralona and Montmauran, all usually viewed as *H. sapiens*. The inclusion of these specimens in *H. erectus* will bias results accordingly.

Rightmire (1981) considers cranial capacity, biauricular diameter (a measure of cranial base breadth), first lower molar breadth and mandibular robustness. Although they vary, there is no consistent patterning with age, and so he concludes that the variation reflects individual differences and/or overall size. However, some workers (e.g. Cronin *et al.*, 1981; Stringer, 1984; Lestrel, 1976; Clausen, 1989) detect a more definite trend of brain enlargement in later *H. erectus*. Of the other characters, M_1 breadth appears to be genuinely stable in hominids generally, not just in *H. erectus*, which reduces its value as an indicator of stasis in the latter species. In any case, conclusions about M_1 breadth and mandibular robustness are both compromised by the difficulties of identifying isolated mandibles as *H. erectus* as opposed to *H. habilis*/early *Homo* sp. or archaic *H. sapiens* (see above). There is a logical difficulty with considering biauricular diameter: as

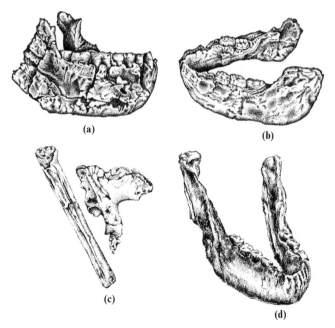

(a)

(b)

(c)

(d)

Figure 7.8 African *H. erectus*. (a) Ternifine 1 mandible; (b) Swartkrans SK 15 mandible; (c) OH 28 hip bone and femur; (d) Baringo BK 67 mandible.

a measure of cranial breadth, its values are largely predetermined by assignment of the specimens to *H. erectus*, one of whose distinguishing features is that maximum cranial breadth is low down near the base, so that the character can be expected to be largely invariant within *H. erectus*.

In later publications, Rightmire (1985; 1986; 1990) reaffirms stasis in *H. erectus* as well as concluding (1985) that 'the study of earlier *Homo* crania casts doubt on the view of human evolution as occurring gradually at a steady tempo throughout the Pleistocene' although the gradualist view neither posits nor requires that the rate of change be constant. A thorough analysis of temporal trends is provided by Wolpoff (1984; 1986), who reviewed patterns of change in 13 cranial characters. Delson (1985) criticised this study on grounds of sample composition, biasing effects, sexual dimorphism and dating uncertainties, criticisms that also apply to Rightmire's studies and, indeed, all other analyses of trends. If the fossil record is too incomplete to document gradualism, it is equally so for stasis. In fact Wolpoff's study identifies clear instances of temporally directed change within *H. erectus*, and in my view undermines assertions of *H. erectus* as a static entity.

7.3 Archaic *Homo sapiens*

Among the later Middle Pleistocene human fossils, including some listed in Table 7.3, are specimens widely regarded as representatives of archaic *H. sapiens* populations. The more complete show numerous similarities with *H. erectus*: cranial capacities may overlap, the vault bones are thick and the skull reinforcement system still moderately well developed; the face is large and projecting, and the supraorbital torus often well developed; palate, mandibular and dental dimensions are all similar to those of *H. erectus*.

However, there are also contrasts: brain size is, on average, rather greater (there is no specimen as small as the smaller *H. erectus* crania), the vault is generally somewhat higher with more inflated frontal and more evenly curved occipital, and the maximum vault width is higher, usually around the level of the temporo-parietal suture, rather than at the base as in *H. erectus*. The nuchal area is still extensive, but the ridge at its perimeter is less clearly defined and it does not extend as far up the occipital, the upper portion of which is near vertical rather than sharply angulated forward, as in *H. erectus*. Sagittal keeling and an angular torus, while sometimes present, are less pronounced than in most *H. erectus*.

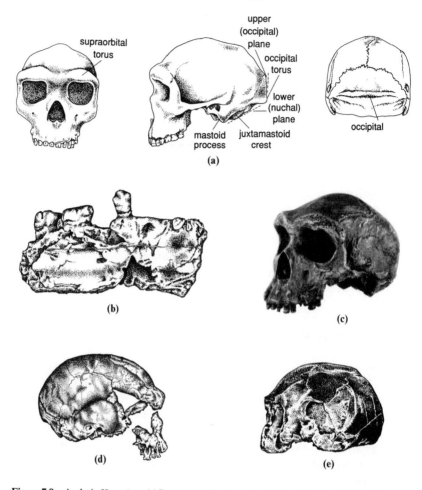

Figure 7.9   Archaic *H. sapiens*: (a) Reconstruction based on Kabwe (Broken Hill) specimen to show main features. African archaic H. sapiens: (b) Cave of Hearths (Makapansgat) mandible; (c) Kabwe (Broken Hill) cranium. (d) Ngaloba (Laetoli: 18) cranium; (e) Eliye Springs cranium.

For many workers the distinction between *H. erectus* and *H. sapiens* is based primarily upon small differences in cranial capacity, neurocranial shape and proportions, while others (e.g. Stringer, 1981; 1984) have sought to identify characters that more precisely define *H. sapiens* as opposed to *H. erectus*. The nature of the distinction is considered further below, after a review of the expanding fossil record for early *H. sapiens*.

7.3.1 Sub-Saharan Africa

Until the early 1970s sub-Saharan Africa was regarded as a relict region, a cul-de-sac for later Quaternary human evolution: archaic morphologies lingered there until late in the Upper Pleistocene, and the pool of cultural/ technological innovation was stagnant. A revised dating framework with a fivefold extension of the evolutionary time-scale and an expanded fossil record have dramatically altered the picture: the region is now seen as being as important for the later stages of human phylogeny as it was for the initial hominid radiation. Far from being a backwater, sub-Saharan Africa appears a focus for phyletic and technological innovation and change over much of the Middle and Upper Pleistocene.

As a result of these developments it is clear that archaic *H. sapiens* specimens from many sites in East and southern Africa (Table 7.3) are not recent but derive from the later Middle Pleistocene. They are usually assigned to *H. sapiens rhodesiensis*, the sample of which has been greatly extended over the last 15 years. Specimens are generally large faced and robustly constructed, some (e.g. Kabwe, Bodo) exceptionally so, and with some similarities to African *H. erectus* (e.g. OH 9). However, cranial capacity is generally greater than in that species ($\geqslant 1200 \, \text{cm}^3$) and the vault bones thinner, although there may be evidence of cranial reinforcement in sagittal keeling and a slight angular torus. Similarly, the nuchal plane, although laterally extensive, does not extend as far up the more vertical occipital, and the cranial base is relatively short and flexed. The face is long and broad with a large (sometimes massive) supraorbital torus, large orbits and nasal aperture, the cheek region inflated and the palate long. Mandibular remains are few (Kapthurin, Baringo, and Cave of Hearths, Makapansgat) and incomplete, but suggest that the lower jaw, whilst robust, was less so than in *H. erectus*.

The more complete of this material is from East Africa: the type specimen from Kabwe, Zambia (formerly Broken Hill, south Rhodesia); LH 18 from the Ngaloba beds, Laetoli; and the recently recovered specimen (ES 11693) from Eliye Springs, Turkana, Kenya. The Saldana (Hopefield) cranium from Cape Province, Florisbad, Orange Free State, and the Cave of Hearths mandible document the species in southern Africa.

The Kabwe material derives from several individuals and besides the cranium includes postcranial elements (humerus, pelvis, femur and tibia). Apart from robustness the postcrania are essentially modern and indicate a relatively tall individual (*c.* 175 cm). The more robust specimens probably represent a male and perhaps go with the cranium; other material is less robust and probably female.

Overall, the specimens probably span a period of perhaps 0.4 my down to just over 0.1 my ago, i.e. around the Middle/Upper Pleistocene boundary. Some phyletic change might thus be expected to be apparent, and Brauer (1984; 1989) has recognised two grades within this material. Archaic *H. sapiens* grade I consists of earlier, more massive specimens such as Bodo (Kalb *et al.*, 1982), Kabwe, Eyasi, Saldana, Ndutu and Cave of Hearths; archaic *H. sapiens* grade II specimens include Omo 2, Laetoli 18, Florisbad and ES 11693. While still stoutly constructed and sharing undoubted similarities with the grade I sample, these specimens show progressive features such as lighter faces, broader, more vertical frontals, laterally expanded parietals, more evenly curved occipitals with a reduced, undercut nuchal area, and diminished supraorbital and occipital tori. While dating of some specimens (e.g. Eliye Springs) is uncertain, the group as a whole is rather later than the grade I fossils, with some specimens perhaps as young as basal Upper Pleistocene (i.e. around 0.12 mya) (see chapter 9).

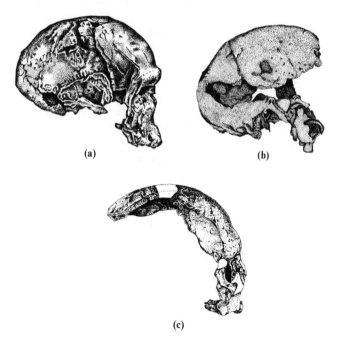

(a)   (b)

(c)

Figure 7.10   African archaic *H. sapiens* (cont'd). (a) Jebel Irhoud 1; (b) Salé cranium; (c) Florisbad.

7.3.2   North Africa

A regional hominid sequence is known from Middle Pleistocene coastal sites in north-west Africa extending from western Algeria around to the Casablanca region on the Atlantic coast of Morocco. The oldest specimens are three mandibles, a parietal and some teeth from Ternifine (Tighenif) near Palikao, Algeria, dated to $c.$ 0.6 mya or earlier on micro- and macro-faunal evidence. The mandibles are large and robust, show similarities with the ZKD specimens and are usually assigned to $H.$ $erectus.$ Other remains also assigned to $H.$ $erectus$ are mandibles from the Littorina cave, Sidi Abderrahman, and from Thomas I Quarry, Casablana, and a partial cranium from Thomas III; all these are rather younger (probably $c.$ 0.4 my old).

About the same age is the braincase from Salé near Rabat, Morocco. This is thick walled and low vaulted, with a retreating frontal, mid-sagittal keeling and a capacity of only 960–1000 cm^3. With these $H.$ $erectus$ features are others that are more reminiscent of $H.$ $sapiens$ so that the specimen is variously attributed to both species. The occipital is strongly curved with only a weak nuchal torus, and the nuchal plane is low set. However, Hublin (1985) has drawn attention to the abnormal pathology of the nuchal region and its effect on the specimen's rear vault morphology, making precise detail difficult to identify. A later $H.$ $sapiens$ specimen (0.15–0.2 my old?) is the partial skull of an adolescent from Kebibit, Rabat, which shows even occipital curvature and a weak nuchal torus, but with a deep face, large dentition and robust, strongly constructed jaws.

The fossils from Jebel Irhoud (Ighoud), Morocco, date from the final Middle or early Upper Pleistocene (circumstances of recovery prevent precise dating) (Grun and Stringer, 1991). Four specimens are known. Irhoud 1, the most complete, is an adult cranium with long vault, retreating frontal and flattened parietals. The supraorbital torus is well developed, the face moderately prognathous and the palate large. Irhoud 2 is the vault of a younger adult, generally similar to 1 but lacking the face and most of the base. Irhoud 3 and 4 are, respectively, a child's mandible and humerus, both strongly built. The specimens were originally identified as Neanderthals but they lack distinctive Neanderthal features (see chapter 8) and are best regarded as advanced archaic $H.$ $sapiens.$

Overall, the limited North African material spans a period of 0.5 my or more and suggests morphological transition rather than abrupt dis-continuity between $H.$ $erectus$ and $H.$ $sapiens$—a pattern continued in the subsequent Upper Pleistocene material from the region (see chapter 8).

7.3.3   *Europe*

Hominid fossils are first known from Europe during the Middle Pleistocene. The great majority of specimens are ≤ 0.4 mya or younger (some
considerably so), but the Heidelberg mandible and a few scattered
archaeological finds suggest at least intermittent human occupation of the
continent during the earlier Middle Pleistocene or even final Lower
Pleistocene.

The Mauer (Heidelberg) mandible was recovered from a gravel pit near
Heidelberg in 1907 and so is among the earliest fossil evidence for human
evolution. It is massively constructed with a thick, internally buttressed,
deep body and low broad ramus, and is among the most prognathous
known. It has often been attributed to *H. erectus*, but the teeth are small and
show moderate pulp enlargement (taurodontism), a trait seen in later
European fossils. There are other detailed differences from the Javan and
Chinese *H. erectus* (Howell, 1960), and it has therefore been argued that
the specimen is more appropriately considered within *H. sapiens*. The
associated fauna is a relatively primitive Middle Pleistocene one of warm
temperate affinities indicating interglacial conditions. No absolute dates
are available, but the specimen could be up to *c.* 0.6 my old.

Other, later and more complete specimens (Arago, Petralona, Swanscombe, Steinheim) provide more definite evidence of early European *H.
sapiens.* Some of these (e.g. Arago, Petralona) are ruggedly built with
strongly constructed supraorbital tori and broad prognathous faces, thick
vault bones with sagittal reinforcement, and a large palate similar in many
respects to *H. erectus*, whereas others (Steinheim, Swanscombe, Fontech-
evade) are more lightly built; such variability may reflect sexual dimorphism. Cranial capacity ranges from a modest 1100–1200 cm^3 (Steinheim,
Petralona) to estimates of more than 1400 cm^3 (Swanscombe, Vertesszollos),
i.e. within to well beyond the *H. erectus* range. Distinctiveness from *H.
erectus* is seen in the vault contour (see above): the lower side walls are more
vertical, the occipital more rounded with a less extensive nuchal scale and a
longer, near-vertical upper scale.

The Petralona specimen is relatively complete, lacking only the mandible. It probably dates from 0.2–0.4 mya, and is perhaps nearer the earlier
rather than the later figure, but more precise dating is not possible. Despite
this uncertainty it provides important evidence of early European *H.
sapiens* morphology. Its small (1200 cm^3) cranial vault is low and thick
walled with marked pneumatisation, a retreating frontal and an extensive
nuchal plane on the projecting occipital. The face is strongly built, broad

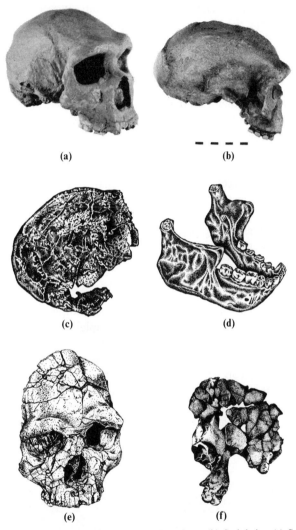

Figure 7.11   European archaic *H. sapiens*. (a) Petralona; (b) Steinheim; (c) Swanscombe; (d) Mauer mandible; (e) Arago face; (f) Arago hip bone.

and prognathous with projecting torus, big nasal aperture and large palate. The less complete Arago 21/47 cranio-facial specimen is similar in many respects with its broad, strongly constructed face and marked alveolar prognathism; estimated cranial capacity is also similar to Petralona.

Steinheim, the other relatively complete Middle Pleistocene cranium, is smaller, and while the brow ridge is still prominent the mid-face is flatter and more lightly constructed, with a distinct canine fossa. The bones of the vault are thinner and the occipital contour more even, with only a weak nuchal torus even in the medial portion. The three Swanscombe bones (both parietals and occipital) are essentially similar in shape, indicating a vault similar to Steinheim's but absolutely larger (c. 1340 cm^3) and thicker.

There are a number of fragmentary and/or immature remains from French sites (Biache, Lazeret, Fontéchevade, Montmaurin, La Chaise) dating from towards the end of the Middle Pleistocene ($\leqslant 0.2$ mya) mainly from cave in-fills, usually indicating cool/cold conditions. Some (e.g. the Montmaurin mandible), although smaller than *H. erectus* and the earlier *H. sapiens* specimens, are still quite strongly built; others (e.g. Biache, La Chaise) are less robust and show features which resemble those of the earliest Upper Pleistocene specimens from western Europe (Stringer, 1985); these are considered in more detail in chapter 8.

7.3.4 *Asia*

The central and east Asian fossil record is far less complete than that further west. In the Indian subcontinent, for example, the only fossil is the Narmada cranium (Sonakia, 1985). This consists of the right side of a skull vault and face with part of the left endocast, associated with a late Middle Pleistocene fauna from Hoshangabad, Madhya Pradesh, central India. It is *H. erectus*-like in general size and proportions, and is probably best considered an advanced form of that species, although the occipital contour appears rather *H. sapiens*-like.

From the Far East the record is still sparse: after the later *H. erectus*

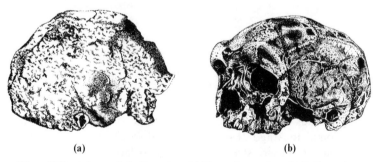

(a)   (b)

Figure 7.12   Asian archaic *H. sapiens*. (a) Narmada cranium; (b) Dali cranium.

crania from ZKD (locus H1) and Hexian there are two important specimens from around the Middle/Upper Pleistocene boundary generally regarded as archaic *H. sapiens*. From Dali in Shaanxi Province, North central China, comes a relatively complete cranium with robustly constructed vault of *c.* 1200 cm^3, pronounced temporal lines and strong straight supraorbital torus. There is continuity with the earlier material in frontal keeling and an angular torus, but the face is remarkably low and flat and the rear vault evenly curved. The less complete Maba specimen is from a cave in Guandong (Canton) Province, south China, and consists of a partial skull cap. It resembles Dali in its sagittal keel, narrow nasal bones and continuous straight supraorbital torus. Sometimes described as an Asian Neanderthal, it differs from that group in many respects and is best thought of as an archaic *H. sapiens*.

7.4   *H. erectus* or *H. sapiens*?

Later Middle Pleistocene specimens from all continents show certain common features so that many of the fossils described above, e.g. Mauer, Vertesszollos, Bilzingsleben, Bodo, Omo, Salé and Ndutu, have been assigned to either *H. erectus* or *H. sapiens* by different workers. There is clearly considerable variability within later Middle Pleistocene material and consequent disagreement about the nature of the *H. erectus/H. sapiens* boundary.

While this disagreement may partly reflect varying perceptions of subjective taxonomic boundaries, it cannot be solely due to this cause since many of the contentious specimens show combinations of features otherwise only found in crania invariably assigned to one or other of the two species. Such intermediacy is powerfully suggestive of continuity in the fossil record, and some workers (e.g. Bilsborough, 1976; 1978; Jelinek, 1978; 1980) have stressed the similarities between specimens and the arbitrary nature of the boundary. Other workers have emphasised the contrasts; Stringer (1978; 1981; 1984; Stringer *et al.*, 1979) in particular, has attempted to clarify a pattern of character distribution between the species using both clade and grade approaches.

So extensive is the overlap that some workers (Howell, 1981; Stringer, 1981) have doubted all evidence for *H. erectus* in Europe, arguing that the claimed *H. erectus* specimens show only features that are also similar in archaic *H. sapiens*; however, one of the latest European specimens, Bilzingsleben, is particularly like *H. erectus* (Stringer, 1984b). Relevant

African specimens include those of Olduvai Bed IV, Ndutu and Bodo. The Bodo 1 specimen, although assigned to *H. sapiens* (Kalb *et al.*, 1982), shows many similarities to Asian *H. erectus*, as does the Bodo 2 parietal (Asfaw, 1983), and the presence of such features in some later (but not earlier) African specimens suggests that characters regarded as typical Asian *H. erectus* may be chronological rather than geographical developments.

In South-East Asia the Ngandong crania, described above as *H. erectus*, have sometimes been considered as archaic *H. sapiens*, although recent studies (especially Santá Luca, 1980) suggest that they are most appropriately regarded as an advanced subspecies of *H. erectus*. Stringer (1984) suggests that similarities between Ngandong and archaic *H. sapiens* may be parallelisms resulting from endocranial expansion, but provides no evidence as to why these, as opposed to resemblances between or within European and African samples, should be regarded as parallel evolution.

The debate, of course, is not just about taxonomic placement *per se*, but whether *H. erectus* and *H. sapiens* are separate clades or merely arbitrary chronospecies within an anagenetic continuum. Some, like Rightmire (1981; 1984), regard *H. erectus* as a 'geographically widespread but relatively homogeneous real species that need not be defined arbitrarily by reference to chronology or gaps in the fossil record, and one that was stable over a long period of Lower and Middle Pleistocene time'. Others see it as merely the arbitrary chronospecies antecedent of *H. sapiens* within a single evolving clade.

The issue is complicated by the lack of a firm chronology, and also by the failure to differentiate adequately between various groups of what are usually termed archaic *H. sapiens*, so obscuring significant patterns of diversity (Tattersall, 1986) (see chapters 8 and 9). Despite such difficulties, on balance the accumulating numbers of morphologically intermediate specimens from all three Old World continents point to continuity between late *H. erectus* and archaic *H. sapiens*. Cladistic techniques cannot identify continuity, but it is significant that their methods have failed to locate any uniquely derived characters of archaic *H. sapiens*, or to provide evidence of a hiatus between the groups. Moreover the repeated, geographically dispersed, occurrence of morphologically mosaic specimens intermediate between *H. erectus* and *H. sapiens* is incompatible with a punctuational interpretation of human evolution.

The regional records of Europe, Africa and Asia provide fossils documenting morphological continuity between *H. erectus* and *H. sapiens* and characterised by broadly similar morphological trends. To invoke parallelism as an explanation for this becomes increasingly unconvincing as the examples multiply. Instead there is a correspondingly increased probability that the separate continental records form part of a larger scale, broadly Old World, anagenetic pattern of morphological change, with little, if any, good evidence to support a Middle Pleistocene speciation event leading to a distinct *H. sapiens* clade.

7.5   Ecology and behaviour of Lower and Middle Pleistocene hominids

Hominid evidence from the greater part of the Lower Pleistocene, both archaeological and fossil, is still confined to sub-Saharan Africa. Some earlier sites, e.g. Koobi Fora, Swartkrans and especially Olduvai, continue to provide important information, but over the period there is an expanding list of additional sites including Chesowanja, Kilombe and Olorgesailie (Kenya), Gadeb, Middle Awash and Melka Kunturé in Ethiopia, and several sites in southern Africa. In many instances the localities are more extensive and contain greater densities of artefacts and bone than final Pliocene/basal Pleistocene sites, suggesting longer periods of occupation and/or larger social groups or both. There is also greater site diversity, with some apparently concerned mainly with artefact production ('factory sites') and others with disarticulation of large carcasses ('butchery sites') as well as less clearly identifiable ('occupation') sites.

In Africa pebble tool industries continue throughout the period, but tend to display greater flaking proficiency and a wider range of forms than the earliest Oldowan. At Olduvai these more advanced, developed Oldowan assemblages occur from around Middle Bed II (*c.* 1.6 mya) onwards. They contain fewer choppers than the typical Oldowan (< 30% compared with up to 80%) and many more spheroids and small scrapers. A few (< 5%) bifacially worked tools, varying in regularity of form, also occur, as do large numbers of small flakes.

Only slightly later (around 1.5 mya) Acheulean assemblages occur at Olduvai and other East African sites. In addition to numerous sharp and retouched flakes, these contain many more bifaces (> 40%) shaped into distinct tool forms, almond-shaped hand axes or transverse-edged cleavers. These tend to be larger, better balanced and more symmetrical than pebble tools, and are more regularly flaked over a greater proportion of the surface. They are often made by retouching large flakes obtained by splitting cobbles, or struck from large boulders, so that one face (the inner face of the original blank) is flatter, with less retouch than the other.

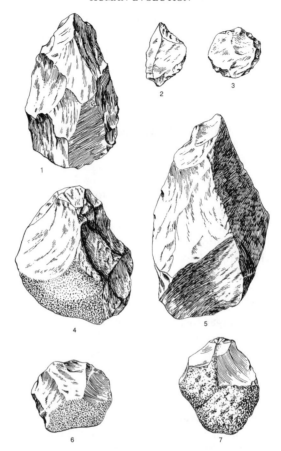

Figure 7.13 Developed Oldowan artifacts (Olduvai). (1) Larva biface; (2) awl; (3) discoid scraper; (4) proto biface; (5) trihedral biface; (6) side chopper; (7) end chopper (from Leakey, 1971).

Basalt, ignimbrite, chert, quartz or obsidian may be carefully selected for some tools, with the source often a considerable distance from the locality where the tools are found. Hand axes at Gadeb, Ethiopia, are struck on obsidian from *c.* 100 km away (Clark and Kurashina, 1979) implying extensive group movement or transport networks: either explanation requires significant curation of handaxes or their blanks. Most flakes result from simple striking with a hammer stone, but by the later Middle Pleistocene they may be produced by the 'proto-Levallois' technique, where several small flakes are removed to create a platform which, when struck,

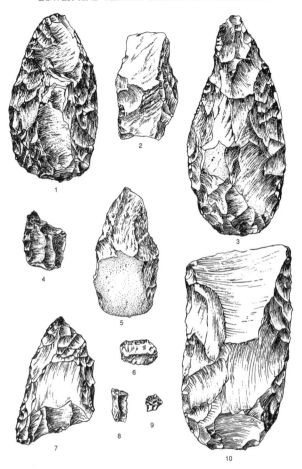

Figure 7.14   Acheulean artifacts (Olduvai and Gadeb). Hand axes: (1,3) Gadeb; (5) Olduvai. Cleaver: (10) Olduvai. Scrapers: (4,7) Gadeb; (6,8,9) Olduvai. (1,3,4,7 from Clark and Kurashina, 1979; 2,6,8,9 from Leakey, 1971).

removes a flake of predetermined size and shape. Retouched and utilised flakes occur in large quantities at Acheulean sites, and it is possible that hand axes originated from the form of the residual core after flakes had been removed for utilisation.

The earliest pebble tool industries predate the Acheulean, but at Olduvai and elsewhere both assemblages exist in Bed II and beyond. This has been interpreted as indicating the existence of two cultural or technological traditions, with the developed Oldowan as 'an uninterrupted local continuation of the Oldowan' (Leakey, 1971). However, the presence of bifaces in the developed Oldowan as well as the Acheulean, the persistence of both for > 1 my and their intercalation at several sites has also led to suggestions that both are functional variants of the same lithic complex. This interpretation is supported by the presence of 'later' Oldowan-like 'pebble tool' industries in North Africa and Europe. It is possible that further work will reveal that several distinct entities have been subsumed under the general labels 'Oldowan' and 'Acheulean'.

The Ethiopian localities at Melka Kunturé and Gadeb on the Highland Plateaux either side of the Rift, and in the Middle Awash, provide additional perspectives. Developed Oldowan and Acheulean floors are present at both Melka Kunturé and Gadeb sites, documenting occupation of the high plains around 1.5 mya or soon after. The locations are close to water which would have supported ribbon forest in an otherwise open environment, and both are large, multiple context, reoccupied sites with evidence of a variety of activities. By contrast the Middle Awash sites are smaller, sparser and represent transient, specific behaviours.

Clark (1987) suggests a pattern for *H. erectus* social organisation in these relatively dry environments based on chimpanzee models, where small groups range widely and may temporarily coalesce with other groups to exploit varied resources on a seasonal cycle. Gadeb and Melka Kunturé represent reoccupied base camps in the riverine forest where most food resources exist, from which smaller groups would forage to outlying areas for specific foods (fruits, tubers, insects, meat, etc.) and for tool materials, activities represented by the Middle Awash sites. Once the resources of a base camp area were temporarily exhausted, the group would rapidly move to reoccupy another adjacent area of gallery forest. In this way a considerable area might be covered in the course of a year, setting in context the distant-sourced Obsidian tools at Gadeb.

The Melka Kunturé localities (Chavaillon *et al.*, 1979) also document longer term shifts: artefacts become more refined with increasing standardisation of bifaces, cleavers and scrapers, greater flake retouch and, in the late Acheulean, Levallois technique. There are also changes in the spatial arrangements, activity patterns and internal organisation of the sites that postdate but parallel the tool kit changes. Acheulean living floors at Melka Kunturé are generally flatter and more regular than those of the Oldowan, and show greater differentiation of activities, with clearly delimited areas for butchery, tool production, etc. By Upper Acheulean times there is evidence of postholes for a shelter, fire and ochre at the site.

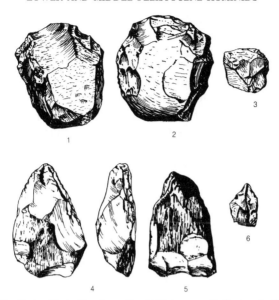

Figure 7.15 Far East artifacts. Zhoukoundian: (1, 2) choppers; (3, 6) retouched flakes. Burma, Anyathian (fossil wood): (4) proto hand axe; (5) hand adze (from Movius, 1948).

By late Lower/early Middle Pleistocene times pebble tool and hand axe assemblages are known from North Africa (Sidi Abderrahmann, Morocco; Ain Hanech and Sidi Zin, Tunisia; Tighenif/Ternifine, Algeria) and Middle East (Ubeidya, Israel). There are also possible occurrences in southern Europe at the Mediterranean sites of Vallonet (France) and Isernia (central Italy), but the overwhelming majority of European finds are < 0.5 my old. Most contain hand axes, but some flake/pebble tool industries (Clactonian, UK; Tayacian, France; Buda, Hungary) are known; their relationship to the widespread European Acheulean is unclear.

The Acheulean complex extends across the Middle East to India, but is not known from the Far East or South-East Asia, where Oldowan-style pebble tool assemblages ('chopper/chopping tool' industries) are found. Factors underlying the hominid radiation out of Africa are obscure, but it would certainly have exposed groups to new environments and selection pressures. For example, southern Asian environments are mainly ones of tropical, monsoon and bamboo forest, not grassland. Artefacts from north Thailand indicate hominid presence by the later Lower Pleistocene (> 0.7 mya) and, as such, broadly accord with *H. erectus* fossils from Lantian and Java much further to the north and south. The sparse archaeological record in Asia, and its scrappy nature, lacking clearly defined forms, has prompted suggestions that stone artefacts were only a small part of the total tool kit, which consisted largely of bamboo, wood, rattan, bone, etc. (Pope and Cronin, 1984). If so, the available stone cores and flakes probably functioned as knives, push planes, spoke shaves, etc. for making the organic tools.

Java was doubtless first occupied and subsequently recolonised during periods of low sea level when the Sunda shelf was largely rain forest and mangrove swamp, so making access difficult: the endemic Javanese fauna is restricted and lacks open-country forms (Pope, 1988). Claims for a Javanese palaeolithic industry, the Pajitanian (Pacitanian), made by *H. erectus*, have been shown to be ill-founded, adding support to the notion of a non-lithic technology for Indonesian *H. erectus*.

Rather more evidence is available from China, but most archaeological sites are poorly dated. Zhoukoudian, the best known, is an exception: evidence indicates the cave was periodically occupied between about 0.5 and 0.25 mya. Besides human remains, the deposits have yielded broken and charred animal bones, ashes, hackberry seeds and many quartz flakes and cores. The cave has therefore been interpreted as a seasonal occupation site, with evidence of the use of fire, systematic hunting and exploitation of plant foods in a distinctly human way.

This scenario has been challenged by Binford and Ho (1985) and Binford and Stone (1986), who dispute the association of many of the elements upon which the reconstruction is based, in particular regular, proficient hunting and the use of fire, although the evidence is convincing for the majority of workers (Pope, 1988). Certainly the occupation levels extend over long periods, and the increased seasonality and temperature variation experienced in such high latitudes resulted in shifting ecological pressures for which improved cultural responses would be critical.

H. erectus endocasts reveal clear left occipital/right frontal asymmetry and further expansion of frontal and temporal lobes, which Holloway and de la Coste-Larymondie (1982) view as evidence of hemispheric dominance for symbolism and spatiovisual integration, i.e. complex cognition of a human pattern.

Leakey and Walker (1989) argue that early *H. erectus* already had an extended period of postnatal brain growth, with its corollary of lengthened infant dependency. The Nariokotome youth's pelvis is transversely narrow, and consideration of OH 28 indicates female pelves would have been similar, since sacral width differs little between the sexes. Estimating *H. erectus* neonate brain size from the adult value using a general primate model where adult brain size is just over twice that of the newborn (Martin, 1990) gives a brain of $c.\,400\,cm^3$ at birth, which would make passage through the narrow pelvis impossible.

Leakey and Walker (1989) therefore conclude that fetal brain growth in *H. erectus* must have been slower, with size at birth less than $400\,cm^3$ to effect safe delivery, and that *H. erectus* had already evolved the distinctively human pattern of extending the fetal growth rate well after birth. This has obvious implications for interpretation of *H. erectus* cognitive abilities and social behaviour, especially mating and procreation, necessary to ensure the survival of infants with an extended dependency phase.

In addition, various workers have attempted to infer early hominid mental abilities and intelligence from tool form and diversity. Oldowan tools are undoubtedly simple and show little or no evidence of 'style'. Nonetheless there are clear indications, most notably in the transport of raw materials, of behaviour that is more complex and requires more planning than that of any modern non-human primates (see chapter 6).

Acheulean assemblages appear to indicate greater mastery of stone-working and the imposition of form; later ones, in particular, have a constancy of shape, balance and proportion that makes them aesthetically pleasing as well as functionally efficient. In addition, certain sites, e.g. Kilombe and Olorgesailie (in Kenya), reveal very large numbers of hand axes made to a constant template. Experiment shows them to be particularly effective for butchering large carcasses, for digging, and for removing tree bark, whilst other core tools such as cleavers are also effective for butchering and for branch severing. Wynn (1985) has argued that whereas Oldowan industries show little if any more ability than that displayed by modern apes, Acheulean assemblages require a degree of conceptualisation more akin to that of modern human intelligence. If correct, this means that many of the mental as well as physical hallmarks of mankind evolved during the Lower and Middle Pleistocene.

CHAPTER EIGHT

UPPER PLEISTOCENE HUMAN EVOLUTION

8.1 Introduction

The Upper Pleistocene differs from earlier periods in several respects. Fossils are more numerous and more complete, so that the primary evidence is denser than for earlier phases of human evolution. Coupled with this is an expanded geographical range, so that specimens are known from most major land areas, although their frequency varies considerably. Chronological and environmental frameworks are also known in greater detail, so promoting more precise analysis of population and environmental relationships.

However, time intervals are shorter and morphological contrasts less marked than with earlier human evolution, whilst polytypic patterning is discernible between regionally based samples. In effect, the fine detail of human evolution is being focused on with individual, local and regional diversity contributing more to the observed variation and overlying phyletic change. It is accordingly difficult to disentangle these components, and a different order of interpretation is required from the broader phylogenies of earlier periods. Observed contrasts in the fossil record are likely to reflect a compound of population variation and replacement, gene flow as well as phyletic change, so that unifactorial interpretations will be inadequate: the biological patterns underlying the fossil record, based as they are upon population interactions over space and time, are likely to be complex.

Since the period sees the appearance of fossils similar to modern humans, a primary focus of enquiry is the origin of anatomically modern man. How, when and where *Homo sapiens sapiens* first appeared, the nature of the change and the subsequent spatial and temporal patterning elsewhere are all issues of lively debate. Another concerns the factors involved in the disappearance of the bigger faced, more robust, archaic human morphologies that dominated until now. Interpreting these changes is complicated by the

Figure 8.1 Some important Upper Pleistocene hominid sites.

fuller fossil record noted above; whilst it is more complete than for earlier periods, the densest records are in peripheral areas such as western Europe, which might be thought unlikely as the source or initial focus of significant changes.

The increased fossil evidence is paralleled by a correspondingly richer and more complex archaeological record, and the Upper Pleistocene marks important behavioural and cultural changes. A major shift is that from Middle Palaeolithic assemblages with a moderate range of tool types made from prepared flakes to the more diversified tool kits of the Upper Palaeolithic, with a greater number of composite tools formed from blade elements struck from prismatic cores. The change is not simply one of flint (or other stone) production: the tools are themselves more standardised and specialised, and there is much greater use of organic material (bone, antler, ivory) in the assemblages. In some areas this technological shift appears broadly to match the morphological changes in the fossil record, with blade technology associated with anatomically modern forms, but there are also important exceptions to this generalisation, which are discussed below. Upper Palaeolithic sites are often deeper and more extensive than earlier sites, suggesting changes in population density, group size, social organisation and subsistence patterns, and in the later Pleistocene there was a remarkable flourishing of art forms—painting, engraving, sculpture and modelling. Whatever arguments and ambiguities may apply to earlier periods, by 20 000 years ago the fossil and archaeological records make clear that these were human communities that were distinctively modern in their sociality, modes of communication, conceptualisation and imagery.

8.2   Chronology and environment

The chronology and environments of the Upper Pleistocene are known in finer detail than those of preceding periods, but even here the record is distinctly patchy. Within Europe the most complete records are from the western seaboard and Mediterranean, with those from south-west France and the low countries especially detailed. Evidence from other parts of the continent is thinner, and the records of Asia and sub-Saharan Africa are particularly patchy with, in many cases, isolated climatic episodes placed in a time grid only by reference to the European record.

Within Europe the base of the Upper Pleistocene is well dated at *c.* 125 000–130 000 years BP, with the onset of the last 'Eemian' interglacial represented by the main marine transgression (high sea level) at stage 5e of the deep sea core record when ice volume was less than at any other time in the Middle/Upper Pleistocene. This warm, truly interglacial, phase appears to have lasted only *c.* 10 000 years and to have terminated by 115 000 years ago or thereabouts. It is followed by a *c.* 40 000 year period (5a–d) when the deep sea record indicates cool conditions overall but with two identifiable phases (b and d) of increased ice volume. The European land record shows corresponding fluctuations: in south-west France the climate appears relatively mild with red deer and bovids at many sites, but there is some evidence of the later of these cold snaps (5b). In north-east France (Grande Pile, Vosges) the pollen profile records deciduous oak forest with two brief but intense cold phases at *c.* 85 000 years and > 100 000 years. A similar pattern is evident in the Low Countries where cold intervals are separated by the Amersfoort, Brorup and Odderade interstadials.

After this last phase, a sudden sharp deterioration in temperature at *c.* 75 000 years and lasting to about 60 000 years marks stage 4 in the deep sea record and the onset of fully glacial cold, dry conditions. There is a brief amelioration around 59 000 years before turning cold again, and the period down to *c.* 13 000 years (i.e. oxygen isotope stages 2 and 3) is predominantly

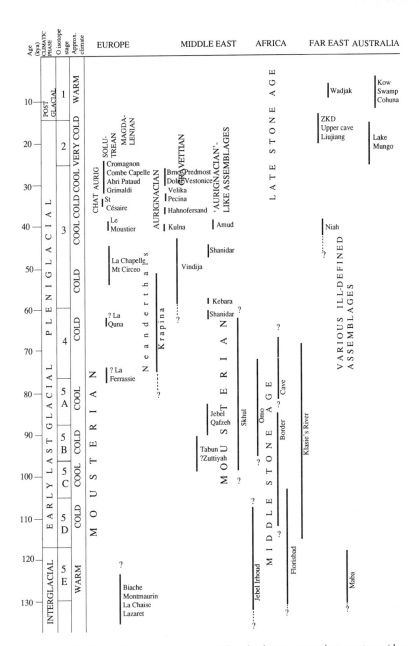

Figure 8.2 The Upper Pleistocene time-scale. Climatic phases, oxygen isotope stages (deep sea core record) and climatic conditions evidenced by the Western European record at left. Probable time ranges of selected hominid sites and artefact assemblages are indicated. Chat. = Chatelperronian; Aurig = Aurignacian. For further details see text.

one of cold, dry polar desert conditions. A short mild phase (Hengelo) offers some respite around 40 000 years, as does a prominent, widely recognised interstadial (Denekamp and equivalents) around 30 000 years, but thereafter conditions again turn cold, with a particularly intense glacial phase around 18 000 years.

Further east the record is less detailed, but at least the broad outlines of the above scheme can be recognised. Here the cold phases are represented by loess deposits of varying thickness, and the warmer episodes by weathering horizons, thicker ones indicating interglacials, thinner ones interstadials between colder periods. Some of the shorter climatic oscillations picked up in the west cannot be recognised in the central and east European record, but the two regions agree in broad outline, especially the onset of the full glacial conditions at around 75 000 years ago.

This traditionally defined the beginning of the last (Würm) glaciation but it is now clear that many sites in France and elsewhere which are considered as 'early' Würm in fact fall within oxygen isotope stages 5a–d (75 000–115 000 years). It is therefore usual to confine the last interglacial to stage 5e, and consider stages 5a–d as within the last glacial, distinguishing this earlier, milder phase from the full glacial conditions or Pleniglacial, after c. 75 000 years.

8.3 Neanderthals and the archaic modern transition

One of the better known groups of *H. sapiens* fossils from the earlier Upper Pleistocene are Neanderthals, extending from the Atlantic seaboard eastwards across Europe and the USSR as far as Uzbekistan, and southwards through the Crimea into the Middle East (Lebanon, Iraq, Israel), and possibly into North Africa. Characteristic Neanderthal remains are known from the last interglacial and the first part of the last glaciation, i.e. 35 000–120 000 years BP, but individual features can be seen in earlier archaic *H. sapiens* specimens, so that the evolution and coalescence of Neanderthal morphology was a relatively slow, drawn-out process. Its disappearance was not: in several areas anatomically modern fossils are found within a very short time of the latest Neanderthal remains, and in some cases even antedating them. The factors underlying this contrast and the interrelationships of Neanderthal and anatomically modern groups represented by the fossils are therefore a major focus in studies of this period. The timing, sequence and pattern of change vary geographically,

Table 8.1 Some important Upper Pleistocene human fossils from Europe and the Middle East.

(A) Neanderthals

Germany
Neanderthal (Feldhofer)	partial skeleton
Saltzgitter Lebenstedt	cranial bones

Belgium
Spy	2 skeletons
La Naulette	mandible
Engis II	child

France
La Chapelle aux Saints	adult skeleton
La Ferrassie	2 adults, 6 children
Le Moustier	youth's skeleton
La Quina	adult, infant
Pech de l'Azé	child's skull
Combe Grenal	fragments of more than 6 individuals
L'Hortus	mandible and other fragments of many individuals
Régourdou	mandible, postcrania
Roc de Marsal	child's skeleton
St Césaire	partial skeleton

Gibraltar
Forbes Quarry	adult skull
Devil's Tower	child's skull

Italy
Mt Circeo	cranium, mandible
Saccopastore	2 crania

Yugoslavia (Croatia)
Krapina	skeletal fragments of many individuals
Vindija	cranial and postcranial fragments of more than 8 individuals

Czechoslovakia
Ganovce	cranial fragments
Kulna	cranial fragments
Ochos	adult mandible, postcrania
Sipka	child's mandible
Sala	cranial fragment

Hungary
Subalyuk	adult mandible, postcrania; child's skull fragments

Russia
Kiik Koba	Adult, infant

Uzbekistan
Teshik Tash	youth

Iraq
Shanidar	9 partial skeletons

(continued)

Table 8.1 (*Contd.*)

Israel
  Zuttiyeh   upper face and frontal
  Tabun (Mt Carmel)   female skeleton, male mandible
  Amud   skeleton, fragments of 3 other individuals
  Kebara   adult, infant skeleton

(B) Early broadly anatomically modern humans

Germany
  Stetten   skull
  Hahnofersand   frontal
  Oberkassel   skeleton

Belgium
  Engis I   adult cranium, fragments

France
  Cromagnon   5 skeletons, additional fragments
  Combe Capelle   skeleton
  Abri Pataud   skull, fragments of other individuals
  Chancelade   skeleton

Italy
  Grimaldi   skeletons: 2 children, 1 adolescent, 3 adults
  Arene Candide   adolescent skeleton

Yugoslavia (Croatia)
  Velika Pecina   frontal

Czechoslovakia
  Brno   skeleton
  Dolni-Vestonice   skeletal fragments of 5–10 individuals
  Mladec   6 skeletons
  Pavlov   male skeleton
  Predmost   skeletal remains of 29 individuals

Austria
  Willendorf   mandible fragment

Hungary
  Istállósko   mandible, teeth

Russia
  Staroselje (?)   infant skeleton

Israel
  Skhul (Mt Carmel)   7 partial skeletons, fragments of 3 others
  Jebel Qafzeh   5 partial skeletons, fragments of more than 10 others

---

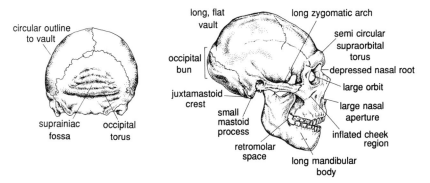

Figure 8.3 Reconstruction of Shanidar 1 to show main Neanderthal features.

and the general treatment below is therefore followed by more detailed regional accounts.

Neanderthal specimens show certain cranial and postcranial features that serve to identify the group as a distinctive variant of the archaic *H. sapiens* morphology. Cranial capacity is medium–large (1200–1800 cm³) and the braincase distinctively shaped. It is long, with a broad retreating frontal, flat parietal region and a projecting occipital forming a 'bun' or 'chignon', although the nuchal area is only moderately extensive. Although not especially low compared with modern *H. sapiens*, the vault may appear so because of its great length and flattened parietals; maximum width is about halfway up the vault so that from the rear it appears almost circular (the '*en bombe*' of French authors).

The elongated vault is a function of the long, relatively unflexed basicranium, as are the projecting face and jaws. These provide one of the most distinctive characteristics of Neanderthals: the face, although not large compared with *H. erectus* or early archaic *H. sapiens*, is set well forward of the braincase and there is considerable midface projection, with the cheek bones both long and inflated with a sloping border to the maxillary process, not high and angled as in modern man. Again, as a consequence of facial projection, the mandibular body is long with a pronounced retromolar space behind M_3 and a sloping ramus with an oblique mandibular angle. The front teeth, set in a broad arc, are sizeable and usually show pronounced wear, while the cheek teeth, although not especially large, often tend to have accessory cusplets and enlarged pulp cavities (taurodontism). Above the jaws the orbits and nasal apertures are large, and the nasal bones sharply projecting, with the root of the nose often depressed. The brow ridges are large and continuous, forming a semi-circular ridge above each orbit. There are other diagnostic features: the mastoid processes are small, the supramastoid crests large, the ear aperture elongated. Despite facial projection the nuchal area is not extensive and there is no distinct torus, but a distinguishing feature is a depression (the suprainiac fossa) on the surface of the occipital just above the superior nuchal line.

Postcranially Neanderthals also show distinctive features. Whilst there is some individual variation, they tend to be short and stockily built with distal limb segments (forearm and lower leg) especially shortened. Limb bone shafts are often curved and the articular joint surface expanded, while the cross-sectional areas of the leg bones and ankle suggest effective resistance to the stresses imposed by a stocky, heavily built trunk, and markings indicate strong tendons for the foot and toe bones for weight support.

In both sexes the pelvis is especially broad and the superior pubic ramus elongated and thinned. This was originally thought to indicate a larger pelvic inlet (and so a larger birth canal in females) but the recently recovered well-preserved Kebara 2 pelvis shows the inlet to be the same size in Neanderthals and modern humans (Rak and Arensburg, 1987; Rak, 1990). Instead the longer pubis reflects a more laterally flared, less tightly curved Neanderthal hip bone, with both sacral and pubic symphysies further forward relative to the hip joint than in modern humans.

On the femoral shaft the linea aspera is typically poorly developed, but other markings suggest a powerful musculature. For example, the shoulder blade of most individuals exhibits a dorsal groove for the powerfully developed teres minor muscle (in most modern individuals confined to the ventral surface of the scapula), which acts by pulling the arm in towards the trunk and rolling it outwards, so complementing the other shoulder muscles that pull the arm down and roll it inward, and so making for fine control as well as great muscular strength. A similar pattern is seen in the hand, with powerful digital flexors and extensors for the finger joints.

All in all, a distinctive and highly particular set of features affects virtually all parts of the skeleton. Many Neanderthals also show limb markings (retroversion of the tibial head, facets on the ankle bones) associated with habitual squatting, but there is no basis for the early influential but now utterly discredited view of Neanderthals as stooped, shambling individuals with imperfect bipedalism (Boule and Vallois, 1957). This picture, above all else the result of Boule's study of the La Chapelle aux Saints skeleton (Boule, 1911–1913) was the result of faulty analysis and insufficient allowance for the deforming effects of osteoarthritis on the trunk and limb skeleton. While the distinctive features of the postcranium are probably related to Neanderthal weight bearing and bodily proportions, and those of the pelvis are particularly intriguing in suggesting minor differences in postural and locomotor biomechanics, there is nothing to indicate an appreciably less developed bipedalism than in other Pleistocene groups.

Available evidence, virtually all from Europe, suggests that the evolution of Neanderthal morphology was a fairly slow, gradual process. The earliest European fossils show few, if any, Neanderthal features, but many of the later Middle Pleistocene specimens display resemblances in one or two features without the complete pattern (Stringer, 1985). For example, the Swanscombe specimen's broad parietals and occipital morphology resemble those of Neanderthals, as does the rear of the much smaller Steinheim vault with its suprainiac fossa. Even more Neanderthal-like are the occipital proportions of the Biache and Saltzgitter Lebenstedt fossils, and the mandibles from La Chaise (Borgeous-Delauney) and Montmaurin with their long bodies and retromolar spaces. All this early evidence is cranial and/or dental: there is little European postcranial material prior to the last interglacial, and what there is (notably the Arago 44 pelvis) is archaic (*H. erectus*-like) but lacks diagnostic Neanderthal features.

By the early Upper Pleistocene, however, a more or less developed Neanderthal morphology is evident in the European record and the Middle East, and with increasing density continues to the first part of the last glaciation. Most of these fossils are the result of burials, often associated with tools and animal bones. These have traditionally been interpreted as grave goods, although some recent studies (Chase and Dibble, 1987) have doubted this, suggesting that they were accidentally commingled at the time of burial. Some have even doubted the act of burial itself, but the evidence appears overwhelming: the consistent flexed position of the body, the presence of several individuals at some sites (La Ferrassie, Spy, La Quina) and the degree of preservation all indicate burial with its social and conceptual implications (see below).

Neanderthal sites are generally restricted in area compared with later occupation of the same locations, and while deposits are often thick overall the individual horizons within are usually thin, suggesting small groups and repeated occupation, perhaps on a seasonal basis. The overall pattern is suggestive of low population density even within those areas where the record is fullest and was presumably so over the entire range of human habitation. Archaeological evidence clearly indicates that tool kits were significantly more complex than those of the Middle Pleistocene. Virtually,

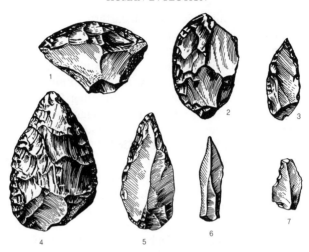

Figure 8.4   Mousterian artifacts. Quina type: (1) transverse scraper; (2) bifacial scraper. Mousterian of Acheulean tradition (hand axe Mousterian): (4) cordiform hand axe; (5) point; (6) backed knife.Typical Mousterian: (3) point. Denticulate Mousterian: (7) denticulate flake (from Bordes, 1961).

but not quite, all Neanderthal fossils are associated with Mousterian industries, characterised by a range of tools, mainly made on prepared flakes. The Mousterian complex has a wide geographical spread from the Atlantic seaboard across Europe and well into Asiatic Russia, northwards to within the Siberian Arctic Circle and southwards to the Himalayan foothills, the Crimea and Middle East and then across into North Africa. It thus extends well beyond the range of known Neanderthal fossils and is also associated, in the Middle East, and (probably) North Africa with some non-Neanderthal fossils.

Over the Mousterian range several major regional forms can be discerned, some largely determined by available raw materials; other, more localised variants, are also recognised, particularly within Europe. The most intensively studied are those of south-west France where Bordes (1961; 1968) has identified several Mousterian facies on the basis of tool type frequencies and manufacturing techniques (Figure 8.4). Their significance is disputed: Bordes (1968) considered them cultural variants made by different groups ('tribes'), Mellars (1970; 1973) favoured a temporal/chronological sequence and Binford argued that they represent functionally differentiated tool kits (Binford and Binford, 1969; Binford, 1973). Elements of all three interpretations are probably involved but the extended time-scale for the Mousterian strengthens the case for broad temporal shifts (Mellars, 1986). At least some tools were complex with several separate elements, and wear patterns indicate differentiated functions.

The sites commonly contain faunal remains, in some cases in large quantities. Binford has argued against systematic hunting, reconstructing Mousterians as scavengers and only opportunistic hunters (i.e. hunting as a result of chance encounters with prey), but the numbers, age structure and

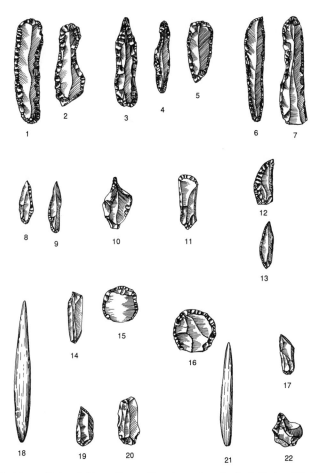

Figure 8.5 Upper Palaeolithic artifacts. Chatelperonian: backed blades (8,9,12,13); end scraper (11); burin (14); thumbnail flake scraper (15). Aurignacian: retouched and strangulated blades (1–7); awl (10); thumbnail scraper (16); burins (17,19,20); bone points (18,21) (from Coles and Higgs, 1969).

restricted species at many sites point to systematic, purposeful hunting, as do the several cliff fall localities known with faunal assemblages at their base.

Mousterian and other Middle Palaeolithic (flake) industries are followed by Upper Palaeolithic assemblages (Sonneville-Bordes, 1963). These typi-

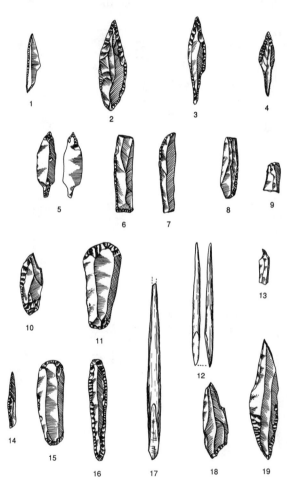

Figure 8.6   Early Upper Palaeolithic artifacts. Gravettian: backed blades (1,14); graver (2); tanged (Font Robert) points (3–5); retouched blades (6–8); burins (9,10); end scrapers (11,13,15); bevelled base bone points (12,17). Proto-Magdalean: retouched blade (16); burins (18,19).

cally contain large numbers of blades, more differentiated tool types, many more composite tools, and appreciable quantities of bone, antler and ivory artefacts. Decorated art objects appeared for the first time in significant numbers, and there are increases in the size, complexity and density of occupation sites. These artefact contrasts have traditionally been taken to imply symbolic, conceptual and social changes that parallel the morphological contrasts between Neanderthal (and other archaic *H. sapiens*) on the one hand and anatomically modern human groups on the other. Excavations over the last 25 years and reassessment of earlier evidence indicate that this is an overly simple picture, and that there is a complex area patterning to the transition. It is therefore appropriate to look more closely at the relevant evidence on a regional basis before attempting an overview.

8.3.1   *Western Europe*

Compared with other areas, the western European fossil record is a relatively full one with a particular concentration of Neanderthal and early modern specimens from south-west France, especially the Dordogne and Charente (Stringer *et al.*, 1984). Many Neanderthal fossils, including the important finds from La Ferrassie (Heim, 1976), La Quina, La Chapelle aux Saints and Le Moustier, were recovered before World War I, and so precise stratigraphic circumstances and dating are often unclear, although recent advances in TL and ESR methods are helping to remedy this (Grun and Stringer, 1991). They are usually associated with a cold fauna and so have usually been dated to the first phase of full glacial conditions (i.e. 70 000–40 000 years ago) but some may date from the preceding cold stages 5b–d. However, there is increased density of both Mousterian sites and Neanderthal remains during the last pleniglacial. ESR results point to the La Chapelle and Mt Circeo (Italy) specimens probably falling within the 55 000–47 000 year span while both TL and ESR techniques indicate the Le Moustier J skeleton to be relatively late at *c*. 40 000 years (Grun and Stringer, 1991; Stringer and Grun, 1991). The latest Neanderthal fossil of all is the Neanderthal partial skeleton from St Césaire (Lévêque and Vandermeersch, 1981) associated with a Chatelperonian (early Upper Palaeolithic) blade industry succeeding the Mousterian levels and TL dated to 36 300 ± 2700 BP (Valladas *et al.*, 1986; Mercier *et al.*, 1991; Stringer and Grun, 1991).

Anatomically modern fossils in the area date from *c*. 34 000–30 000 years ago, only a short period, if at all, after the latest Neanderthals. The first to be

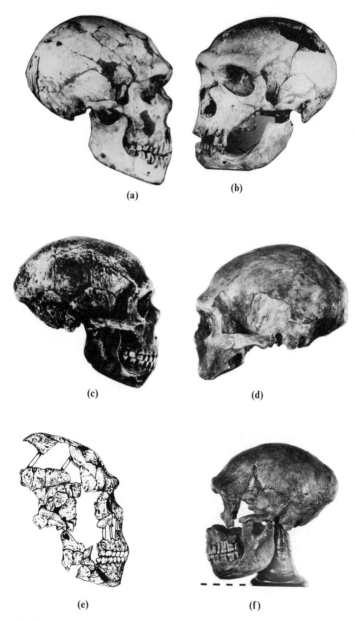

Figure 8.7 West European Neanderthals. (a) La Ferrassie I; (b) La Chapelle aux Saints (unrestored); (c) La Chapelle aux Saints (restored); (d) Mt Circeo I; (e) St Césaire; (f) La Quina.

recovered, and the best known, are those from the rock shelter at Cromagnon, Les Eyzies, which yielded remains of five individuals and other fragments, as well as much archaeological material. Other sites include Combe Capelle (a male skeleton since destroyed), Grotte des Enfants at Monaco (six skeletons, four adults and two children) and Abri Pataud. All these specimens contrast markedly with the preceding material. Cromagnon 1, for example, has a large rounded braincase, a broad short and flat face with no brow ridge and short mandible with a definite chin. Some specimens display a degree of supraorbital development and occipital expansion, but none could be described as Neanderthal, and postcranial differences are similarly marked. All have essentially modern trunk and limbs without any trace of Neanderthal features, and the great majority are tall, with long straight limbs. The specimens are invariably associated with Aurignacian assemblages, a more widely distributed and persistent Upper

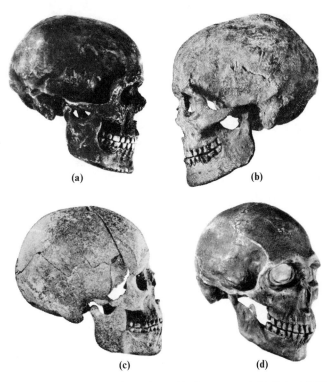

(a)   (b)

(c)   (d)

Figure 8.8   European early broadly moderns. (a) Cromagnon 1 (b) Grimaldi (Grotte des Enfants); (c) Combe Capelle; (d) Predmost III.

Palaeolithic industry than the localised Chatelperonian, and known in south-west France from *c*. 34 000 BP onwards. Other, later, specimens document the continuation of *H. sapiens sapiens* morphology through the later Pleistocene of western Europe with its subsequent technological traditions.

8.3.2 *Central and eastern Europe and Russia*

Although the record here is sparser than that of western Europe, Neanderthal remains are known from a large area extending from east

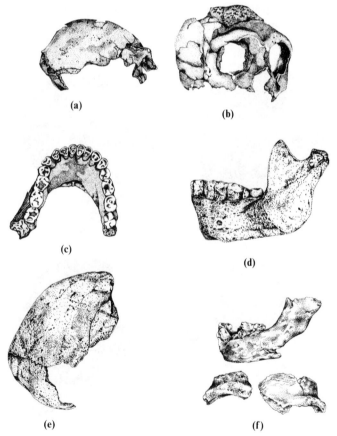

(a)

(b)

(c)

(d)

(e)   (f)

Figure 8.9   South-east European Neanderthals: Krapina fossils. (a) K.6; (b) K.3; (c) K.58 (note breadth of anterior part of dental arch); (d) K 59; (e) K.2; (f) Hortus (France) mandible above and Vindija fossils Cr 262 (below left) and Cr 261 (below right) (from Wolpoff, 1980).

Germany through into central and eastern Europe, and southern regions of the former USSR (Smith, 1984). The largest sample (c. 800 fragments of at least 13 individuals of all ages) is that from the Krapina rock shelter, Croatia, north-west Yugoslavia, recovered between 1899 and 1905, and dating from the end of the last interglacial/early last glacial. Most parts of the skeleton and dentition are represented, although in fragmentary form: the remains are characteristically Neanderthal. Also from Croatia are fragments (eight individuals) from level G3 Vindija Cave. These, although Neanderthal, show interesting contrasts with Krapina: the supraorbital torus is slender and divided above the orbit, and the mandibular symphysis and frontal are vertically orientated. Further east several Neanderthals, including children, are known from the Crimea (Kiik Koba), but the most easterly find is the youth (c. 12 years) from Teshik Tash in the foothills of the Hindu Kush, Uzbekistan, deliberately buried and surrounded by goat skulls.

Early *H. sapiens sapiens* fossils include several important finds from Aurignacian deposits at Mladec (Lautsch), Moravia, recovered between 1880 and 1904, but destroyed in World War II. They differ considerably in size and robustness, perhaps because of sexual dimorphism, and are undoubtedly modern in overall morphology although one specimen (5) showed brow ridge development reminiscent of Neanderthals. Much less complete, but carbon-14 dated to $> 33\,850 \pm 120$ years BP is an anatomically modern frontal from Velika Pecina, Yugoslavia. A rather more complete specimen with slight brow ridge development is the similarly early ($36\,300 \pm 600$ years BP) frontal from Hahnofersand, west Germany.

More complete finds are known from rather later deposits associated with early Gravettian tool kits. They include specimens from the Czechoslovakian sites at Brno (a burial covered with mammoth tusks), Predmost (28 individuals), Dolni Vestonice (5–10 individuals) and Pavlov (a male skeleton); the last three sites are well dated to around 25 000–26 000 years BP. Male specimens in this sample are relatively tall with robust crania. Some display slight brow ridge development but the torus is divided as in more recent crania, and contrasts with the continuous Neanderthal pattern. Again, some crania display slight occipital bunning, but of a non-Neanderthal form.

In summary the central and eastern European evidence indicates Neanderthal groups in late Eem and early Würm down to c. 38 000–40 000 years BP (Kulna), with anatomically modern individuals in Aurignacian contexts from c. 34 000–36 000 years BP until about 30 000 years BP; more extensive Gravettian finds are dated to $\geqslant 25\,000$ years BP by carbon-14. The admittedly sparse fossil record is, however, indicative of a transition in at least some characters (brow ridge development, frontal orientation, facial height, dental and mandibular proportions) between Neanderthal and modern morphologies. As such it is more suggestive of continuity between the populations represented by these specimens than those shown to be identical according to the record in western Europe. The implications of these contrasts are further discussed below.

8.3.3 Middle East

Several Middle Eastern sites have revealed Neanderthal fossils, the most important being the Zuttiyeh and Amud caves, the Tabun and Kabara caves (Mt Carmel), Israel, and the fine series of remains (nine individuals) from Shanidar, Iraq (Trinkaus, 1983a; 1984b; Arensburg, 1989; Vandermeersch, 1989). The oldest dated specimens (by ESR) are the female skeleton and isolated male mandible from Tabun upper level C (or possibly the overlying level B) dated at c. 95 000–100 000 years ago, while the Zuttiyeh frontal and upper facial fragment is probably also early and shows resemblances to some of the early European specimens. Other specimens are similar to one another and, in their broad features, to the bulk of the European last glacial Neanderthals. They differ from these in some details, i.e. orbits and nasal aperture are smaller on average, the nasal root is not so depressed, there is a less inflated zygomatic region, occipital bunning is not so pronounced, and at least some individuals are rather taller and longer limbed than most of the European specimens. Overall, however, their morphology is unmistakeably Neanderthal. Most are probably 40 000–80 000 years old. For example, ESR and TL methods suggest an age of c. 60 000 years for Kebara while the Shanidar specimens divide into two groups: an earlier set around 60 000+ years and a later (more than 45 000 years) set of three specimens. Shanidar I suffered extensive injury during life and his survival has implications for social relationships and cooperation between Neanderthals (see below). The latest set, possibly as young as 40 000 years ago (Grun and Stringer, 1991) appears to be those specimens from Amud, including the skeleton of a tall, large-brained (1800 cm^3) male (Suzuki and Takai, 1970).

Also from the Middle East are robust but broadly anatomically modern specimens, contrasting with the Neanderthals both cranially and postcranially. The first examples were the remains of ten individuals (infants to adults) excavated from Levalloiso-Mousterian deposits (layer B) at the Skhul rock shelter, Mt Carmel, by Garrod (Garrod and Bate, 1937; McCown and Keith, 1939) and by Neuville and Stekelis at Jebel Qafzeh.

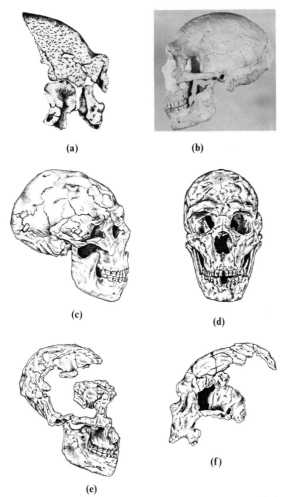

Figure 8.10 Middle East Neanderthals. (a) Zuttiyeh frontal; (b) Amud I; (c–d) Shanidar I; (e) Shanidar 2; (f) Shanidar 5.

More recently Vandermeersch (1981) has recovered further material from Jebel Qafzeh with precise stratigraphic detail, so expanding the sample and demonstrating beyond doubt that the specimens derive from the Levalloiso-Mousterian deposits. Originally thought to be very early (last interglacial), detailed faunal comparisons with Tabun then led workers to suggest that Skhul and Qafzeh were around 40 000 years old. However,

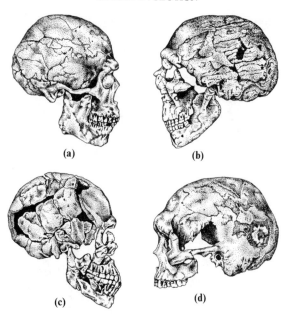

Figure 8.11   Middle East early broadly moderns. (a) Skhul 5; (b) Skhul 4; (c) Jebel Qafzeh 9; (d) Jebel Qafzeh 6.

recent TL dating indicates an age of 92 000 ± 5000 years for the Jebel Qafzeh remains (Valladas *et al.*, 1988). ESR dates agree, although with a rather wider range, and the same technique suggests Skhul is of comparable age (101 000 ± 12 000 years old) or somewhat younger (81 000 ± 15 000 years old) (Stringer *et al.*, 1989; Grun and Stringer, 1991). Both sites therefore predate the bulk of Neanderthal specimens, with important phyletic implications (see below).

The Skhul and Qafzeh individuals are generally tall and long-limbed, with a modern postcranial morphology, high rounded braincase and large cranial capacity. While some are modern and flat faced, exhibiting a definite chin, others have larger, robust and more prognathous faces with a developed supraorbital torus, but even here there are contrasts—the structure and pattern of facial projection is non-Neanderthal, and there is no inflated zygomatic region.

So the fossil record points to contrasting patterns in the three regions: in the Middle East an early appearance of broadly modern morphology and apparent coexistence for *c.* 40 000 years with fully developed Neanderthals, but without convincing evidence of transitional/intermediate groups; in central and eastern Europe suggestions of a gradual/clinal shift from Neanderthal to modern morphology; and nearer the Atlantic coast the persistence of Neanderthals down to at least c. 36 000 years BP, whereafter they disappear, to be replaced only a little later by anatomically modern humans. The sequence apparent in the above points to a pattern of change from central to more peripheral regions, and archaeological data help fill out this interpretation.

In the Middle East a very few sites indicate a transient and precocious occurrence of blade industries within the long-established Mousterian sequence and more widespread Upper Palaeolithic assemblages broadly resembling the European Aurignacian are known from c. 40 000 years BP onwards. Detailed excavation (Marks, 1983) suggests that the shift from flakes to blades occurred in some cases over a very short time-scale. Here Neanderthals and the earliest broadly modern fossils (Qafzeh and Skhul) are firmly associated with Mousterian contexts, whereas in Europe both occur in the Upper Palaeolithic: in addition to Mousterian associations the latest Neanderthals (St Cesaire and Arcy-sur-Cure) are from Chatelperonian levels, whereas H. s. sapiens remains are from Aurignacian deposits, from which there are no Neanderthal fossils.

The relationship of these tool kits to the underlying industries also differs. The Chatelperonian clearly develops from the antecedent Mousterian, and there are similar developments elsewhere in Europe, the Uluzzian in the north and toe of Italy and the Szelethian of the central Danube basin, pointing to regional continuity in the shift from Middle to Upper Palaeolithic (Mellars, 1989). However, the distinctive nature of the Aurignacian, its sudden fully formed appearance and lack of obvious precursors in the underlying assemblages of western Europe (the best-documented region) all point to an exotic origin and intrusion into that area. Given the fossil associations it is most plausibly interpreted as reflecting movement of new, anatomically modern groups into the region, with the Chatelperonian an acculturation response by indigenous Neanderthals/Mousterians to the new techniques and forms of tool production. These were presumably filtering westward by sporadic, intermittent contact before the first pronounced intrusion of Aurignacian groups into the region which would have accelerated the process of copying and innovation. However, the pattern was certainly more complex than just a simple, slow westward trickle. Accelerator carbon-14 dating indicates that at some Spanish sites the Aurignacian replaces the Mousterian as early as 40 000 years ago without any hiatus or intervening Chatelperonian (Bischoff et al., 1988; Straus, 1989). It may be that westward movement of Aurignacian groups along the Mediterranean coast was appreciably more rapid than inland, where prolonged, direct contact with Mousterians was delayed, but where transient encounters stimulated Chatelperonian development.

Further east, a transitional or wholly modern morphology may well predate the central European Aurignacian, as it does in the Middle East. The transitional finds summarised above (Vindija, etc.) would then represent *in situ* development as a result of gene flow from the Balkans or other areas of south-east Europe or (perhaps less likely) hybridisation between Neanderthals and anatomically modern groups. The assumption here is that the period of change resulting from gene flow and sporadic contact would precede the intrusion of modern morphology on any significant, discernible scale as a result of group movement or expansion, itself probably a consequence of the ecological and behavioural changes reflected in the archaeological record. Certainly the density, complexity and range of Aurignacian sites is greater than that of the Mousterian or indigenous Upper Palaeolithic variants. Gamble (1986) reviews the European archaeological evidence and sets it within an environmental context.

This interpretation implies the decoupling of modern anatomy from Upper Palaeolithic behavioural/technological developments. Their inter-linkage has been promoted for so long that dissociation is difficult, but it definitely does *not* occur in the Middle East, and even in western Europe is not invariably true. The very earliest *H. s. sapiens* groups were probably subsisting and behaving on a Middle Palaeolithic/Mousterian basis, with the behavioural and conceptual developments associated with the Upper Palaeolithic occurring only later. Modern morphology may, in some ill-defined way, have been a prerequisite for these, but it was hardly an immediate cause. Indeed there is good evidence for increasing social complexity and population density within the Upper Palaeolithic itself, indicating gradual cumulative changes rather than a quantum leap at the Middle/Upper Palaeolithic boundary. Klein (1989) provides a survey of these technological, ecological and behavioural shifts which is valuable because its geographical range extends beyond the traditional areas of Europe and the Middle East.

8.3.4 Far East

The central and east Asian fossil record is far less complete than that of the circum-Mediterranean and western Russia. In the Indian subcontinent, for

Table 8.2 Some important Upper Pleistocene human fossils from East Asia, Australia and Africa.

Far East and Australia

(A) Archaic H. sapiens

China
    (?) Dali   cranium
    (?) Maba   partial cranium

(B) Early broadly modern *H. sapiens*

China
    Zhoukoudian (Upper Cave)   crania
    Liaujiang   cranium

Java
    Wadjak   crania

Borneo
    Niah   cranium

Australia
    Lake Mungo   partial skeleton, skeleton, fragments,
    Kow Swamp and Cohuna   crania and skeleton remains of 40–50
                             individuals
    Cossack   skeleton

Africa

(A) Archaic *H. sapiens*

Morocco
    (?) Jebel Irhoud   skull, cranial vault, other fragments

Libya
    Haua Fteah   partial mandibles

Ethiopia
    (?) Omo 2   cranium

South Africa
    Floris bad   partial skull

(B) Early broadly modern *H. sapiens*

Morocco
    Dar es Soltan   adult skull, child skull, mandible, teeth

Sudan
    Singa   cranium

Ethiopia
    Omo I   partial skull

South Africa
    Border Cave   4 individuals (3 adult, 1 child)
    Die Kelders Cave   teeth
    Equus Cave   mandible and teeth
    Klasies River Mouth   fragments of more than 5 individuals

---

example, there is little or nothing after the *c.* 200 000-year-old Narmada cranium, although archaeological evidence reveals Upper Pleistocene occupation. In the Far East the record is hardly better: there is little after the archaic *H. sapiens* Dali and Maba specimens (see chapter 7) which are Middle Pleistocene or, at youngest, basal Upper Pleistocene in date.

Anatomically modern remains are unknown before the final Pleistocene, the best known being those from the upper cave at Zhoukoudian, dated at around 15 000 years BP. They are ruggedly built with flat, broad upper faces and sagittal keeling but are fully modern, and similar to a specimen of similar age from Liaujiang in Kwangswi province. This last is also ruggedly built, but shows traits in common with contemporary Asiatic populations in its shovel-shaped incisors and lack of a third molar.

In South-East Asia the record is equally sparse, but that from Australia has expanded considerably in recent years. After the Ngandong fossils the only specimens from Java are the Wadjak crania, recovered in the 1880s but not described until much later. They are associated with an essentially modern fauna and are at best final Pleistocene, and possibly even Holocene in date. Whilst morphologically modern they are relatively large and robust, more reminiscent of Australian aboriginal crania than contemporary Indonesian.

To the north of Java is Borneo where the remains of an individual of *c.* 12 years was recovered from the Niah cave. The skull is completely modern in morphology and lightly built with a rounded vault and short flat face. Dating is problematic: a carbon-14 age of *c.* 40 000 years is based on only a single estimate from charcoal in another part of the cave, and so has been widely dismissed. However, Niah's apparent early age is entirely consistent with the expanding Australian record, which has dramatically altered views on the human colonisation of the continent.

Archaeological and fossil evidence indicates that Australia and Papua New Guinea were occupied by at least 35 000 years BP, and probably as early as 40 000–45 000 years BP. Such early dates are remarkable, for the minimal sea level of the last glaciation was not until well after 32 000 years BP when the Molucca Strait crossing between Sulawasi and Papua New Guinea was at least 40 miles; before that the distance must have been even greater. The earliest human groups in the continent must therefore have been able to construct craft capable of making a sea voyage of 50 miles or more. Additional support is provided by the pattern of evidence within Australia: many of the earlier sites are in the Murray Darling basin, in the south-east of the continent, suggesting voyaging along the coast and then entry into the interior along the major river systems. There is no evidence of a general southwards occupation of the continent from Papua New Guinea and the northern coast of Australia.

While many tool kits were relatively simple, their makers were far from being the primitive, impoverished groups popularly supposed; on the contrary, early Australian communities displayed several important cultural and technological innovations such as cremation, rock art, cooperative fishing involving the construction of creeks, dams and large nets, and the use of groundstone axes.

The earliest fossil remains are those of the Willandra Lakes system of the Lachlan River, New South Wales, where there are some numerous archaeological sites with hearths, tools and in some cases human remains. The best known are from the Walls of China Lunette (Dune) bordering Lake Mungo (Barbetti and Allen, 1972; Bowler *et al.*, 1972). Mungo I, dated at 24 500–26 500 years BP, is the partial skeleton of a female who had been cremated, probably quite a long time after death. The individual was lightly built and completely modern with a flat face, only slight brow ridge development and a vertical forehead, but with some frontal narrowing and a broad nuchal area. Mungo III, a more complete skeleton of a male who

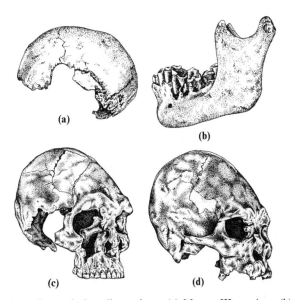

(a)

(b)

(c)   (d)

Figure 8.12 Australian early broadly moderns. (a) Mungo III cranium; (b) Mungo III mandible; (c) Kow Swamp 1; (d) Kow Swamp 5 ((a), (b) from Day, 1986; (c), (d) from Thorne and Wolpoff, 1981).

suffered from arthritis and marked anterior dental wear, is dated to 28 000–32 000 years BP. Like Mungo I the specimen is lightly built, more so, in fact, than most modern Australian aboriginals.

There are other specimens known from south-east Australia; some, including Mossgail, Nitchie and Talgai, are undated but almost certainly later than the above specimens. The best-known later material is that from Kow Swamp near Gumbar Creek, Victoria, dated between 9000 and 13 000 years BP, i.e. final Pleistocene/early Holocene (Thorne and Macumber, 1982). In this area there are numerous crania associated with the old Lake Kow shoreline or buried in the levees of the stream system feeding the lake. Most remains were excavated in 1968–72, but one specimen, the Cohuna cranium, found in the area in 1925 is a generally included with the Kow Swamp sample.

The crania are large and ruggedly built, more so than most Australian aboriginals, and markedly more so than the early Lake Mungo specimens. Faces and jaws are large and projecting with pronounced masticatory muscles and strongly built brow ridges, which in some cases form a distinct torus. Whilst the parietals and occipital region of the vault are high, the frontal is flat and retreating, the vault bones are thick and there may be an occipital torus. These features have been interpreted by some as links with *H. erectus* and evidence for evolutionary continuity in the South-East Asia/Australia region.

However the '*H. erectus*' features are essentially those of a large face/masticatory apparatus superimposed upon a modern morphology, whilst the retreating frontal and associated brow ridge development are likely consequences of artificial skull deformation by head pressing (Brown, 1981). Although by far the largest group, Kow Swamp specimens were not the only such crania known: a similar specimen of about the same age is known from the Cossack cave on the north-west coast of Western Australia, and indicates that the morphology was not a localised regional one.

The contrast between the early, more lightly built, Lake Mungo specimens and the robust Kow Swamp material has prompted speculation about the relative roles of population movement, hybridisation and selection pressures within Australia during the later Pleistocene. In particular, do the two morphologies represent two 'waves' of colonisation of the continent or *in situ* changes in response to selection pressures in Australia? There is insufficient evidence to resolve these issues (Habgood, 1989) but whatever the explanation(s) it seems clear that the contrasts are superimposed upon essentially modern cranial morphology. As such, they provide some indication of the magnitude of anatomical differences that might be expected over a relatively short time interval and a single land mass within the span of a single subspecies, i.e. without significant phyletic change.

Australian and European fossil evidence thus indicates *H. sapiens sapiens* in both areas at the periphery of the Old World by ⩾ 30 000 years BP. With only the disputed Niah specimen from South-East Asia, there is insufficient evidence to detail the initial occurrence of modern groups in areas adjacent to Australia, but the European record is far more complete. The pattern indicates anatomically modern forms present earlier in eastern than western Europe, and earlier still in the Middle East. It might therefore be expected that such groups are present even earlier in more central land masses. The Asian record, its western rim apart, is far too sparse to be informative, but the African record is more complete, and it is from that continent that the more complete evidence for early *H. s. sapiens* is known.

8.3.5   *Africa*

One or two archaic *H. sapiens* crania from both sub-Saharan and North Africa may date from around the Mid/Upper Pleistocene boundary, or even slightly later (see chapter 7). They possibly include Omo 2 (see below), Laetoli 18, Eliye Springs and Florisbad from the east and south of the continent and, from North Africa, the Haua Fteah (Libya) partial mandible and the Jebel Irhoud (Morocco) fossils. However, the most significant development has been the recovery of anatomically modern specimens from sub-Saharan Middle Stone Age (MSA) horizons early in the Upper Pleistocene. Three sites are especially important: Omo (Kibish) 1, Ethiopia; Border Cave; and Klasie's River mouth, South Africa.

8.3.5.1   *Omo (Kibish) 1.*   In 1969 Leakey recovered a partial skeleton from the KHS site in the Kibish sediments of the lower Omo (Day 1972; Day and Stringer, 1982). The specimen was from the near the base of member 1, much lower than the upper part of member 3 which is carbon-14 dated to ⩾ 31 000 years BP. According to Butzer, the entire Kibish sequence (members 1–3) correlates with oxygen isotope stage 5 (i.e. ⩾ 70 000 years BP). A thorium–uranium date of 130 000 years was obtained on freshwater shells from the deposit, but its reliability has been questioned. The Omo 1 skull is incomplete and lacks most of the face, but there are part of the vault, the maxilla, and the lower part of the mandible; the postcranial remains (shoulder, arm, parts of the vertebral column, leg and foot bones) are

essentially modern. The skull is ruggedly built, but the braincase is high and
rounded, with expanded, bossed parietals, and a rounded occipital with
restricted nuchal area and only slight medial development of the torus.
The frontal is vertical, and, while there is some thickening of the brow ridge

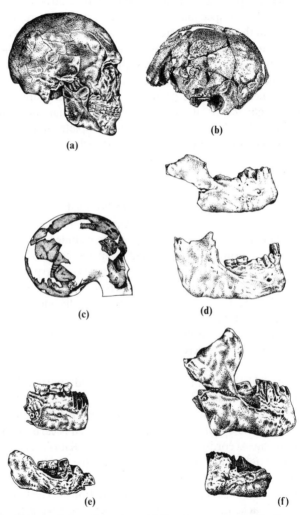

Figure 8.13   African early broadly moderns. (a) Omo I as reconstructed by Day and Stringer;
(b) Omo II; (c) Border Cave 1; (d) Border Cave 2 (above) and Border Cave 5 (below); (e) Klasie's
River Mouth 13400 (above) and 16424 (below); (f) Klasie's River Mouth 41815 (above) and
21776 (below) (from Brauer, 1984).

above the nose, it thins laterally, and the arches are of robust modern rather than archaic form. The mandible shows a clear external chin.

Another specimen (a complete cranial vault, Omo 2) was recovered from site PHS about 2.5 km away. The specimen is decidedly more archaic than Omo 1 and, as a surface find, is insecurely dated. It is generally considered to derive from the Kibish deposits and to be approximately contemporaneous with Omo 1; if so it perhaps represents the latest known occurrence of archaic *H. sapiens* in East Africa (see also p. 168).

8.3.5.2 *Border Cave.* This site, also known as Ingwavuma, is in northern Natal, South Africa, close to the border with Swaziland. Over the last 50 years it has yielded the remains of five individuals, initially from commercial (guano) digging, latterly from archaeological excavations by Beaumont. BC 1 is the partial cranium (with some leg fragments) of an adult found in 1940–42; BC 2 is an adult mandible; BC 3 an infant skeleton (recovered 1942); BC 4 an iron age burial; and BC 5 a mandible found in 1974. Uncertainties remain with BC 1–3 but BC 5's provenance is more secure. While some argue it may be intrusive, there are pointers to considerable age: Beaumont's excavations have clarified stratigraphy at the site, and Butzer has identified 15 sedimentary units within the long sequence of MSA industries dated by intersite correlation and climatic reconstruction between 49 000 and 190 000 years ago. BC 1 and 2 are correlated with level 10 on the basis of matrix adhering to the specimens, the BC 3 infant from levels 9/10 112 000–115 000 years BP, and BC 5 was excavated from level 8, considered to be around 90 000 years ago. Recent ESR dating reduces these estimates somewhat. Grun and Stringer (1991) argue that BC 1 and 2 are < 90 000 years old, BC 3 is 70 000–80 000 years old and BC 5 probably 50 000–65 000 years old.

All the remains are anatomically modern: BC 1 is fairly ruggedly built, with some slight brow ridge development, but is otherwise unremarkable; the BC 2 and 5 mandibles have small borders and well-developed chins. The infant skeleton corresponds closely with that of a modern child of about the same age (3–6 months). Some workers consider BC 1 to be most like recent South African crania (Khoi, San or Nguni) (Rightmire, 1979a; 1981; De Villiers and Fatti, 1982). However, a more recent analysis (Van Vark *et al.*, 1989) suggests that, while modern, the specimen has no especially close or direct similarity with recent southern African populations.

8.3.5.3 *Klasie's River.* Caves and rock shelters around the mouth of Klasie's River, on the Cape coast of Plettenburg Bay have been excavated by Singer and Wymer (1982) yielding extensive archaeological deposits with fragmentary human fossils. The sites are associated with the 6–8 m beach usually correlated to the last interglacial, and oxygen isotope analysis places the earliest MSA horizons within stage 5e, i.e. 120000–130000 years BP. The deposits are extensive and the site was evidently occupied for a period of 50 000 years or more since the latest MSA horizons are c. 60 000 years old or even less. The human remains represent several individuals and are mainly derived from levels MSA 1 and 2, estimated to be between 120000–80000 years BP. Although fragmentary, preservation is sufficient to indicate anatomically modern morphology. For example, a frontal fragment from cave 1 shows no trace of a torus and the superciliary arches are of modern appearance; several mandible fragments, although varying in size, are generally modern in their proportions with a vertical internal contour to the symphysis and a clear external chin. Although Binford (1984) on the basis of his interpretation of the fauna challenges the chronological sequence of Singer and Weimer (1982), re-excavation by Deacon and Deacon evidently confirms the original interpretation, as does the palaeo-temperature analysis. ESR dates (Grun and Stringer, 1991) suggest that the latest MSA deposits may be only c. 40000 years old but confirm the antiquity of the early MSA horizons, with the earliest human fossils > 90000 years ago and others between 60000 and 85000 years ago. The remains from Klasie's River, although fragmentary, attest to the early presence of modern humans at the southern tip of Africa.

8.3.5.4 *General.* The above three sites provide the best evidence for early anatomically modern *H. sapiens* in sub-Saharan Africa. The case is certainly not incontestable: the early dates for Border Cave and Omo are less certain than Klasie's River and even this is not as secure as, say, the dates of many European sites. However, this is not unexpected given the much fuller record from Europe and the fact that the sub-Saharan sites are beyond carbon-14 range. The convergence of different lines of evidence, all pointing to similar dates in the early Upper Pleistocene, is impressive, and provides the firmest evidence, along with the Middle East sites of Skhul and Qafzeh, for the early occurrence of anatomically modern *H. sapiens.*

8.3.6 *The Americas*

All evidence points to human entry into the New World only during the late Pleistocene, when the low sea levels of the last glaciation connected Asia and America across the Beringia land bridge. Mousterian and Upper

Palaeolithic hunters exploited the megafauna of the cold, dry Siberian tundra, especially mammoth, woolly rhinoceros and bison, and there are scattered archaeological sites along the Yenisey, Lena and Aldan Rivers, and around Lake Baikal.

Beringia, which served as a migration route for caribou and other species, was exposed by falling sea levels between 26 000 and 11 000 years BP, and it is perhaps likely that hunters were present in Alaska by *c.* 20 000 years BP, although actual evidence is no older than *c.* 10 000 years, and even then is extremely scanty. In any case, the extensive North American (Wisconsin) ice sheets blocked access to the continental interior until the terminal Pleistocene, so that Beringia/Alaska was, in effect, an easterly extension of the Siberian biome. Glacial melting during a brief warm interval around 12 000 years BP opened a corridor along the Mackenzie River between the Rocky Mountain and Laurentian ice sheets and so provided access to the high plains of continental North America. Thereafter archaeological evidence indicates a rapid spread of hunting groups, even down to the tip of South America by *c.* 11 000 years BP.

The earliest assemblages contain scrapers, gravers, knives and fluted lanceolate points (Clovis points), sometimes associated with mammoth bones. As mammoth numbers declined there was a shift in hunting strategies indicated by increased bison remains and broadly similar tool kits but with smaller, more finely made Folsom points.

Archaeological evidence is sparse, probably reflecting the extreme mobility of these hunting groups, and skeletal remains virtually unknown, but there is no suggestion of any hominid other than *H. sapiens sapiens* in the Americas. However, debate continues as to whether there was human occupation of both Alaska and continental North America before the earliest reliably dated sites towards the very end of the Pleistocene.

8.4 Contemporary evidence

Analysis of current human diversity has also been brought to bear on the recent evolution of *Homo sapiens*. In addition to the extensive data on serological and other polymorphic systems accumulated over the last half-century, recent developments in molecular biology have provided more detailed information on patterns of genetic variation in human populations. These confirm classical morphological and serological surveys in demonstrating remarkably high intra-group variability compared with inter-group diversity: for most characters between-group differences are swamped by the magnitude of the variability within groups. For example, Nei and Roychoudhury (1982) used data from 62 protein and 23 blood group loci of Caucasoid, Negroid and Mongoloid groups to investigate heterozygosity and polymorphic loci (i.e. those significantly above mutational frequency), and concluded that $c.90\%$ of the total variation was within- rather than between-group variability. Wainscoat et al. (1986) examined a more restricted genetic system, five linked loci in the beta globin gene complex from a wider range of populations. Despite the high intra-group diversity, both Nei's and Wainscoat et al.'s studies point to greater separation between African and non-African populations than between any pairs of non-African groups.

Because of its genetic properties, attention has been particularly directed towards analysing patterns of mitochondrial DNA (mt DNA) in human groups. Most DNA is in the cell nucleus, contributed by both parents. Prior to gamete production recombination occurs, mixing the maternally and paternally derived DNA in new combinations within the sperm and ova, so making it impossible to track molecular lineages in nuclear material. However, some DNA is also found in mitochondria, the cells' respiratory centres, as well as in the nucleus. The short length of mt DNA does not recombine and is only inherited through the mother, so that its study in principle allows tracking of maternal lineages.

Cann et al. (1987) sequenced mt DNA from 147 individuals from five major continental areas: [sub-Saharan Africa, Asia, Australia, Papua New Guinea; and Caucasia (Europe, North Africa and Middle East)]. They found that the 147 mt DNAs sorted into 133 types, but only seven of these are from more than one individual. Of the groups, Africans are the most variable, followed by Asians, with the analysis again indicating that most diversity is within groups rather than between groups, so that most mt DNA variation is shared between populations.

Cann et al. (1988) then constructed a phyletic tree for the 133 mt DNA types; the most parsimonious interpretation (i.e. the shortest tree) has a primary split leading to two branches: one exclusively African (but not including all African types), the other leading, via subsequent splittings, to all five populations, including the other African types. Cann et al. thus conclude that Africa is the likely source of human mt DNA, since this origin minimises the number of intercontinental migrations needed to account for the observed mt DNA diversity. The tree includes more than one mt DNA type in each region, implying multiple colonisations of each continent: 27 in Asia and 36 in Europe as minima. Other trees, including one with primary branches leading exclusively to each of the five populations, involve considerably more mutational steps, and are therefore rejected on the parsimony principle (Cann, 1988).

The minimum length tree can be given a time-scale by assuming a constant rate of molecular evolution and using archaeologically derived estimates of primary colonisation events for initial calibration. These indicate a mean rate of mt DNA divergence of 2–4% per million years, giving estimates of 140 000–290 000 years for the common ancestor of modern human mt DNA and 112 000–225 000 years for the primary split. Migration out of Africa may be as recently as 23 000 years or as much as 180 000 years ago; the figure of 2–4% per million years for the rate of mt DNA divergence is compatible with estimates derived from dating divergences in other species, and so is considered reasonable by Cann *et al.* Vigilant *et al.* (1991) further analyse mt DNA data indicating an African origin for *H. s. sapiens c.* 166–249 000 years ago (but see pp. 218–219 and 233–234).

The case for an African origin for maternally inherited mt DNA has led to descriptions of 'Eve' and an African 'Garden of Eden'. However, images of a single founding mother for humanity are wide of the mark: what is described is a single mt DNA type that may have been present in a population of several hundreds or thousands of females, most of whom left no mt DNA descendants simply because of chance. Given the small group sizes, small sibships and high mortality likely to have characterised Pleistocene groups and a sex ratio of 1:1 where males never transmit their mt DNA type, it is highly probable that mitrochondrial diversity would be progressively reduced to a single type solely as a result of random factors.

8.5 Interpretations

Contemporary genetic diversity and the available fossil evidence thus both point to Africa as the likely focus for the initial appearance of modern humans, but is such a pattern more likely to be artefact than reality? What of the selection pressures and evolutionary processes involved in the appearance and establishment of *H. s. sapiens*? Did anatomically modern humans originate with a speciation event and so were reproductively isolated from other Upper Pleistocene *Homo*, and then expanded outwards, supplanting and replacing more archaic morphologies? Or did modern morphology evolve as a result of anagenetic processes, i.e. a gradual reconstruction of the entire gene pool of later Pleistocene *H. sapiens*? This would eventually be across the entire span of the species, but at varying rates and to differing extents: in central areas gene combinations determining modern morphology might well be at appreciably higher frequencies because of the greater magnitude of gene flow than in more peripheral regions, and the early African fossils might simply reflect this phenomenon.

High intra-group genetic diversity and the absence of genetic traits uniquely confined to particular populations has long been regarded as evidence of extensive gene flow and thus, by implication, support for the anagenetic model. In principle this can be viewed as composed of vertical threads representing continuity over time, and horizontal threads representing gene flow over space, combining to form a reticulate, or network, pattern. Such an interpretation has been labelled the regional continuity model, although there are, in fact, several forms of the model differing principally in the time depth accorded to anagenesis and the importance given to gene flow or its absence, i.e. isolation. For example an early, now largely discredited, version was that of Weidenreich (1947) and Coon (1962), who viewed modern populations as evolving largely independently from precursor *H. erectus* but with minimal gene flow between regions, so stressing local continuity combined with isolation back as far as the early Middle Pleistocene. In contrast, Dobzhansky (1962) argued powerfully from the perspective of evolutionary genetics for the importance of gene flow in later human evolution. Another variant of the model, associated, for example, with Thorne (1981) and Thorne and Wolpoff (1981), stresses regional continuity, especially at the periphery. In this version regional variation should first become distinct at the species limit, with local selective factors acting upon the original, limited, founder gene pool, and the resulting features are maintained by a balance between gene flow and selection that results in morphoclines. In this way regional contrasts persist, yet selectively advantageous alleles spread across the entire species range. The emphasis therefore is on local continuity *and* the reconstruction of the gene pool as a whole (the assimilation model (Smith *et al.*, 1989)) but not on the precocious appearance of modern morphology in any particular region; indeed Wolpoff (1989) has questioned the early dating of African *H. s. sapiens* sites such as at Klasie's River.

Brauer (1984; 1989) has proposed a geographically limited 'hybridisation and replacement' model to account for the fossil evidence from Africa, the Middle East and Europe, which he terms the 'Afro-European *sapiens* hypothesis'. This sees anatomically modern *H. sapiens* evolving anagenetically from more archaic forms in Africa (see pp. 209–212) and then expanding beyond the continent as a result of climatic and environmental change. Such expansion might have occurred quite slowly, and resulted in variable gene flow, in part owing to a 'ripple effect' so that the Skhul and Jebel Qafzeh fossils on this model represent hybrid populations which in turn hybridise with other archaic *H. sapiens* groups, e.g. in south-east Europe. The process is similar to that envisaged by Bilsborough (1983) in being multicausal and regionally variable, and the resulting patterns accordingly complex.

Stringer and Andrews (1988) have recently compared the multiregional model of recent human evolution with a single origin 'Noah's Ark' model (Howells, 1976) originating in Africa by speciation. Both models can accommodate the early appearance of anatomically modern forms in Africa (see above); the major contrasts are in evolution beyond Africa, and the processes involved. The anagenetic, regional continuity model views subsequent evolution outside Africa as partly the result of *in situ* evolution, greatly accelerated by gene flow, with or without migration.

By contrast, the single-origin model posits speciation and initial regional differentiation of anatomically modern humans within Africa, followed by their expansion beyond the continent *without* extensive genetic contact with the indigenous populations whom they replace, and finally the recent establishment of the non-African regional characteristics of modern populations. The high intra-group, low inter-group variability of modern humans would thus simply be a reflection of the shallow depth of the modern human gene pool in many parts of the world, rather than a consequence of long-established patterns of gene flow of appreciable magnitude. On this model there has simply not been time for inter-population differences to accumulate, save possibly between African and peripheral populations. Stringer and Andrews (1988) construct predictions for both models, and consider that the fossil and modern genetic evidence both accord better with the recent African origin for anatomically modern *H. sapiens* by speciation, rather than with the multiregional model and its emphasis on anagenesis.

They cite particularly the high intra-group genetic diversity of African populations, the African/non-African differentiation revealed by mt and nuclear DNA, and the age estimates for the origin and divergence of modern mt DNA assuming neutrality and a constant evolution rate. Within the fossil evidence they identify a basic east-west division among Middle Pleistocene hominids, replaced by one of greater similarity across Eurasia in the late Middle/Upper Pleistocene, so eroding regional continuity. Discontinuities between European Neanderthals and their contemporaries in the Far East and Australasia imply a lack of clinal patterns, and therefore gene flow, whilst the hiatus between Neanderthals and anatomically modern forms, the latter's early appearance in Africa and then the Middle East, and the likely coexistence of both morphologies there for *c.* 50 000 years all militate against the regional model. The similarity of early *H. s. sapiens* crania from Europe, North and sub-Saharan Africa, Asia and ·Australia further implies a lack of regional differentiation and points instead to basic 'African' features distributed on an Old World basis.

However, interpretations of both kinds of evidence by Stringer and Andrew (1988) have been severely criticised. Spuhler (1988) argues that the high regional variation of mt DNA and the estimated duration of the variants favour a regional continuity model. In particular, the presence of unique variants in Asia and Europe estimated to be > 200 000 years old supports the notion of regional continuity at least as far back as archaic *H. sapiens*, and a revised rate of mt DNA evolution (0.5–1% per million years) extends its time of origin back to 400 000 years ago, i.e. to *H. erectus*. The present pattern of human mt DNA variations (if confirmed by larger samples) may thus reflect the original expansion of *H. erectus* out of Africa (*c.* 1 mya?) rather than the origin of anatomically modern humans. Even this becomes problematic with the recent identification of shorter trees with a non-African origin (see pp. 233–234). The value of mt DNA for tracking human origins is far from clear.

Similarly, the fossil evidence is less than wholly clear cut. While the western European record points to discontinuity and replacement, that of central and eastern Europe is more suggestive of continuity. Chinese workers have long argued for continuity between Middle Pleistocene and modern populations in the Far East on the basis of cranial and dental traits, while Wolpoff *et al.* (1984) and Wolpoff (1985) have strongly argued for continuity in South-East Asia and Australia. Stringer and Andrews (1988) consider aspects of the diversity of the Australian fossils problematic for both models, but Habgood (1985) provides evidence of strong regional grouping of Australasian crania. In the western part of the Old World, the variability of the Skhul and Jebel Qafzeh remains and of the European Upper Palaeolithic fossils may be taken to accord with the recent African origin model. However, archaic features of the Middle Eastern remains (supraorbital torus, large faces, some occipital proportions) are suggestive of continuity, as is the central European record (see above).

Overall, the record reveals complex patterning, suggesting replacement in some areas, continuity in others (Jelinek, 1985; Wolpoff *et al.*, 1988; Smith *et al.*, 1989). Such a mix, indicating demic (local population) persistence or replacement, is quite compatible with continuity on a larger, continental scale. On the other hand, the 'out of Africa' speciation model requires the reproductive isolation and total replacement of *all*, African and non-African, archaic populations by anatomically modern humans. It is difficult to reconcile such a fate with both the patterning of the available fossil evidence and the ecological and behavioural flexibility indicated by the archaeological record (Smith *et al.*, 1989).

8.6 Selection

The timing and nature of the archaic–modern transition was spatially diverse and morphologically complex. Most detailed analyses and speculations about selection pressures have focused on Neanderthal–modern contrasts; there have been few attempts to consider other archaic *H. sapiens* in any detail. This is partly for historical reasons, particularly the Eurocentric bias of much earlier palaeoanthropology, and partly because Neanderthals make up the bulk of the fossil evidence, and their chronology is relatively secure.

Even with a relatively full fossil record conclusions depend upon assumptions about the heritability of morphological traits. The genetic basis of skeletal variation is unknown but certainly complex, and there is increasing awareness of the influence of environmental factors, e.g. biomechanical forces, in shaping skeletal form. A significant degree of phenotypic lability in many contrasts either side of the transition would allow rapid morphological change over short time intervals, and so would be much easier to reconcile with, for example, the western European evidence than interpretations which suppose a predominantly genetic component to the observed changes. However, the relative contributions of genetic and environmental factors to most skeletal variation remains frustratingly obscure. Moreover, it is difficult to identify obvious functional advantages for some anatomically modern features, while their ubiquity makes it impossible to tie them into specific environmental factors.

The changes to modern morphology, whilst often rapid, were not abrupt: Trinkaus and Smith (1985) draw attention to detailed mosaicism and stress the need to calibrate velocities, distinguishing between features that are relatively stable, those that change without peaking at the transition, those showing peak velocities at that time, and those where shifts are confined to the transition. Trinkaus (1983b) and Trinkaus and Smith (1985) also state the interactive nature of many behavioural and morphological shifts, so that teasing out the two is a complex and largely artificial exercise.

Neanderthal features were traditionally interpreted as reflecting isolation in Europe and cold adaptation to the rigours of the last glaciation (e.g. Howell, 1952; Coon, 1962). There are difficulties with this, not least because

Neanderthals were not confined to Europe, were not isolated there, and their morphology is found prior to the last glaciation. Moreover, the facial features claimed as cold adaptations fail to achieve an internally consistent complex and contrast markedly with those of modern Arctic populations, and so are particularly unconvincing. However, there is no doubt that certain aspects of Neanderthal morphology and proportions—the large robust and compact body mass and relatively short limbs, especially their distal segments—would be adaptive in cold-stressed environments and predicted on thermoregulatory grounds. As such they would represent long-standing adaptations to high latitudes since Middle Pleistocene fossils (e.g. Zhoukoudian) show similar proportions.

Other features of Neanderthal postcrania indicate a well-developed musculature and powerful yet finely controlled manipulation. By contrast, early modern postcrania have weaker muscle markings, and are more linear and gracile, with especially long lower limb segments suggestive of hot environments (Trinkaus, 1983b). Such proportions are seen in both the European Upper Palaeolithic fossils and the much earlier Jebel Qafzeh/ Skhul individuals who coexisted with Neanderthals for > 30 000 years. Trinkaus argues that their distribution reflects cultural mechanisms for thermoregulation in cold habitats, so allowing tropical bodily proportions to spread beyond the tropics. However, this cannot account for the origin of modern morphology, and there is no independent evidence of such cultural improvements, since the earlier phases of the expansion are associated with tool kits little or no different from those of Neanderthals (see above).

An alternative interpretation of Neanderthal cranio-facial morphology is that it represents a biomechanical response to forces resulting from the use of the front teeth as tools for gripping, softening materials, etc. ('paramasticatory' activity). On this view archaic *H. sapiens* (not just Neanderthal) faces are determined by the need to resist powerful biting forces, especially in the anterior teeth; their large size and heavy wear provide some support for this view, as do certain other features of the Neanderthal face (Smith, 1983) although detailed biomechanical models and mechanisms differ: compare Rak (1986), Demes (1987) and Trinkaus (1987). Other aspects of skull morphology such as the long, flat vault, anteriorly positioned face, supraorbital torus and occipital bunning may also reflect anterior dental loading. Flat-faced, modern *H. sapiens* are inferred not to have used their teeth and jaws in this way, although there is no direct evidence that their initial tool kits facilitated a wider range of activities than those of Neanderthals.

Genetic and developmental aspects are particularly relevant to consider-ations of cranio-facial form. It is comparatively easy to envisage the selective advantages for a large face and powerful jaws, but more difficult to see what general advantage might accrue from small structures that is sufficient to explain the rapid, ubiquitous and sustained reduction in facial size. Brace (1963) posited a much criticised 'probable mutation effect' where, once selection favouring larger faces was relaxed, the accumulation of neutral mutations would lead to a reduction in facial size. This does not follow and, in any case, does not adequately explain either the appearance of small faces or the disappearance of large ones. Others have suggested that once a larger face is no longer needed the energetic cost of maintaining it will result in smaller faced individuals having a selective advantage (Smith, 1983). Again, this 'general economy' argument is hardly a convincing explanation for the scale and pace of change, and the virtual disappearance of large-faced forms.

Alternatively, if facial size and proportions are at least partly a phenotypic response to biting stresses experienced during development, much more rapid change can be accommodated. There is some evidence that teeth are under strong genetic control but face and jaw proportions are more labile, so that rapid morphological contrasts do not necessarily imply equivalent evolutionary (i.e. genetic) change. Even so, the shifts in jaw forces must have been virtually universal to produce the observed contrasts in the fossil record, while the slow initial evolution of Neanderthal facial features implies a strong genetic basis for many of these.

Another explanation for the facial and masticatory contrasts is that once selection favouring large jaws is relaxed modern features reflect basicranial remodelling associated with vocalisation rather than food processing. Lieberman (Lieberman et al., 1972; Lieberman, 1975) argued that Nean-derthals' language capability was only c. 10% that of modern humans, without the ability to produce vowels a, i, o, u and consonents g and k. This conclusion was based on reconstructing the Neanderthal vocal tract from surviving basicranial evidence, comparisons with modern humans and ape throats, and computer simulations of sounds based on the vocal tract reconstruction. The approach has been much criticised at each level of analysis: the fragmentary nature of the Neanderthal basicranium, its relevance for vocal tract reconstruction (Seigal and Carlisle, 1974; Burr, 1976) and the vocalisation model derived (Wind, 1978). More recently, recovery of the Kebara Neanderthal's hyoid bone (Arensburg et al., 1989), the first to be preserved, shows that while face and jaws differ from those of modern humans the hyoid is virtually identical. Its features indicate a throat position similar to that of modern humans, and contrasts markedly with Lieberman's reconstruction from which the vocalisation range is derived. There is thus no good evidence to indicate that Neanderthal vocalisation was less extensive than our own.

From the early nineteenth century onwards many workers have pointed out that modern human cranial features (the large globular braincase, small face and jaws) are neotenous, that is they are 'infant' features persisting through into adulthood. Trinkaus (1983b; 1984a) has ingeniously combined these with other postcranial contrasts to argue that there were differences in gestation length between Neanderthals and moderns. The broad Neanderthal pelvis indicated by the extended public ramus would allow a bigger brained infant to pass through and permit Neanderthal mothers to carry babies for 11–12 months before birth (around the time predicted from the average primate relationship between gestation length and adult brain size). Reducing gestation length to the 9 months of modern humans means a less mature neonate but a longer period of exposure to diverse environmental stimuli when the central nervous system is rapidly developing. Provided maternal behaviour and cultural mechanisms allow the survival of such less-developed infants, the behavioural advantages following from exposure, as well as the greater locomotor efficiency from a narrower pelvis, would, in Trinkaus's view, be sufficient to establish such a developmental pattern.

The suggestion is a fascinating one, but there is no independent evidence of either longer gestation length in Neanderthals or of the greater behavioural complexity necessary for neonatal survival in the earliest anatomically modern forms. Most telling of all, the Kebara pelvis indicates that despite contrasts in morphology, pelvic inlet size was similar in Neanderthals and moderns (see above). Indeed, Leakey and Walker (1989) suggest that the characteristically human, non-typical primate developmental pattern is of much greater antiquity and was established by 1.5 mya (see chapter 7). The causal factors underlying the shift to moderns remain tantalisingly obscure.

8.7 *Homo sapiens sapiens*

The archaic–modern transition marks the most recent major shift in the hominid fossil record, but the period since has not been one of stasis. Late Pleistocene and Holocene human samples show contrasts with each other and with modern humans that, while less dramatic than those summarised above, are impressive in their consistency. The general picture is one of decreasing size and robustness, especially of the face, jaws and dentition, but also of stature.

This is most clearly seen in Europe, where Frayer (1977; 1984) has analysed contrasts between early, mid and late Upper Palaeolithic and Mesolithic (post-Pleistocene) samples. In many dental dimensions the early and mid-Upper Palaeolithic groups resemble Neanderthals more closely than they do late Upper Palaeolithic/Mesolithic specimens and there are contrasts in facial projection, vault size and shape and stature, with the later groups shorter and more gracile than the earlier samples. Reduction is more marked in males than females, implying different selection pressures on the two sexes, and a progressive reduction in the degree of sexual dimorphism. A similar reduction in cranial robustness and tooth size is seen in some Asian populations (Brace et al., 1984) and in late Pleistocene/early Holocene samples from South and East Africa.

In this sense the term 'anatomically modern' applied to Pleistocene specimens is, if interpreted literally, something of a misnomer since they may well fall beyond contemporary human morphological variation (e.g. Howells, 1973). 'Broadly anatomically modern' would be a better, albeit imprecise, label.

Wolpoff (1980), Frayer (1984) and others have interpreted these changes as further exemplifying the interplay of behaviour and morphology, especially those resulting from dietary shifts and changes in food pre-paration techniques. Similar difficulties apply as with interpretation of earlier changes, and the precise mechanisms are correspondingly obscure. Nonetheless they certainly document rapid and significant changes in skeletal morphology and so help place the earlier transition(s) in context. Other similar instances are the generally smaller face and jaws of early agriculturists compared with hunter–gatherer groups and, most recently, the trend towards increased height and weight in many populations over the last 200 years or so.

Such changes doubtless reflect biomechanical, nutritional or develop-mental (endocrine) shifts and as such are largely, perhaps wholly, non-genetic, except in the very general sense that there is a genetic basis permitting such morphological flexibility. They exemplify rapid, short-term changes that could not be sustained over long periods, for the transformations would then be so marked as to be functionally or developmentally disruptive and are therefore simply not picked up before the final Pleistocene. Similar influences, however, almost certainly contributed to the shift from archaic to modern *H. sapiens,* and above all else these instances illustrate the plasticity of skeletal morphology that needs to be accommodated in phyletic judgements.

CHAPTER NINE

HUMAN EVOLUTION: PATTERNS, PROBLEMS AND PROSPECTS

The preceding chapters briefly survey palaeoanthropology, draw attention to gaps as well as certainties, and point out the distinction between the 'hard' evidence of the fossils and contextual data and the interpretations based upon that evidence. The account, as with all such attempts, is inevitably subjective since it is influenced by a particular conceptual framework. Even basic descriptions of the fossils follow from many judgements about features that are more or less relevant, while the interpretive passages obviously reflect the author's assessments of the evidence and other workers' views.

The approach is adaptationist with the minimum number of clades judged to be compatible with the evidence. While viewing early hominid diversity as largely polyphyletic (a shift of perspective compared with, say, a decade ago), a significant role is also attributed to anagenesis (phyletic evolution), especially over the last 1.5 my or so. Overall, the account represents a broadly gradualist interpretation of hominid evolution, and a stratophenetic approach has been used where appropriate, notably in chapters 6 and 7. There are certainly other ways of interpreting the evidence. Groves (1989), drawing upon speciation theory and his wide studies of mammalian systematics, provides a stimulating review of hominid phylogeny from a mainly punctuational perspective, deriving a correspondingly greater array (16) of early hominid forms. By contrast, Tobias (1991) sets *H. habilis* within an evolutionary framework dominated by anagenesis and with reticulate evolution as the primary mode over the last million years.

Recognition of the role that investigator bias inevitably plays in all palaeoanthropology is among the more important conceptual advances made within the subject over the last decade. Given such subjectivity and the range of conflicting interpretations, can any general statements be made about the pattern of human evolution? Are any long-term trends evident?

This chapter attempts to summarise features of the fossil record, discuss some of the conceptual issues involved in current controversies, and suggest likely areas of future research. Because of compression, the personal element is particularly prominent with most assertions unsubstantiated, although the relevant evidence and any qualifications can generally be found in the earlier, more detailed, sections.

A major problem of paleoanthropology, for some *the* problem, continues to be the how, when and where of hominid origins. It is clear that the late Tertiary ape radiation was extensive and highly varied, so that by the later mid-Miocene (12–10 my BP) there are multiple fossils from Europe, the Middle East and Asia, associated with diverse environments. Some specimens show dental traits (e.g. enlarged molars, thick enamel, small canines) in which they resemble hominids, although one of the best-known groups (*Sivapithecus/Ramapithecus*) is now firmly established within the orang clade, and so can be dismissed as a human ancestor. With others it is not clear whether the hominid-like features indicate close phyletic affinity, are merely primitive retentions (plesiomorphies) or reflect sexual dimorphism, as with the small canines of some specimens. Many of the features cited as evidence of hominid status for the newly recovered skull of *Ouranopithecus* (de Bonis *et al.*, 1990), which has undoubted dental resemblances with some early *Australopithecus* specimens, may be of this kind (Andrews, 1990).

A variety of data indicate chimp and gorilla to be closely related to humans, but as yet there is only scrappy and incomplete fossil evidence from the African later Miocene and Pliocene. The discovery of fossils to document the common ancestry of African pongids and hominids, and the evolution of the chimp and gorilla lineages, about which next to nothing is known, remains one of palaeoanthropology's goals. Such finds should also resolve which of the two, chimp or gorilla, is more closely related to humans: current studies conflict on this issue, some pointing to one, some to the other, and others to a trifurcation. Part of the problem is the very small number of extant hominoid species, which makes the character polarities used in cladistic analyses uncertain; an expanding fossil record should help to refine these patterns, so increasing the value of modern data as well as being of major importance in its own right.

Recovery of African fossils in the 5–10 my period should also yield clues to the initial evolution of two crucial functional complexes, the masticatory apparatus and dentition, and the locomotor system. The relationship of the latter to arboreal clambering and knuckle walking is an open issue, as is the possibility of several bipedal modes. The absence of fossils makes it unclear whether bipedalism evolved once or several times, the environmental circumstances which elicited such locomotor behaviour(s), and relationships to other locomotor patterns. Postcranial remains of the immediate common ancestor of hominids/African pongids (if recognised as such), and of early members of the individual lineages, should help to resolve these uncertainties.

These remains should also clarify aspects of the early hominid radiation, for australopithecine evolution, traditionally viewed as a two-pronged affair—graciles leading to *Homo*, robusts leading to nowhere but extinction—is now seen as considerably more complex. The Hadar and Laetoli fossils recast schemes of hominid phylogeny, but raised as many issues as they resolved. If *Australopithecus afarensis* is a single species, then the range of variation, especially postcranial, is very great, indicating dimorphism at least equal to or greater than that of the largest extant primates; if more than one species is represented at Hadar/Laetoli then the phyletic pattern is even more complex than currently appears. Further perspectives on this problem are provided by Stw 252, noticeably more macrodont and prognathic than other Sterkfontein member 4 specimens, and claimed as evidence for a second australopithecine species at the site (Clarke, 1988). This interpretation is unlikely, but the specimen both extends the intraspecific range of *A. africanus*, and erodes some of the claimed differences between that species and *A. afarensis*.

In fact, the scale and pattern of sexual dimorphism, its impact on within-species variation, and thus between-species differences, has become a theme, explicit or implied, of many studies. Hadar and Sterkfontein aside, it is clear that previous notions of limited australopithecine dimorphism, based on analogy with modern humans, are untenable. The expanded samples now available show appreciable body size variation at all major sites, and so greater intraspecific ranges and reduced between-species differences.

There is also evidence for marked sexual dimorphism in some earlier Tertiary hominoids at least, with claims that the Lufeng fossils, for example, represent not two species as previously thought, but one with a degree of variation exceeding that of gorillas and orangs (Kelley and Xu, 1991). This raises real problems of hypothesis testing and interpretation (Martin, 1991) but, if correct, suggests early hominids may not have been unusual in their degree of dimorphism but may instead have exemplified a more general Tertiary large hominoid pattern. Reduced dimorphism, characteristic of modern hominids, would then be a late development (?*H. erectus* onwards) which raises interesting issues about the adaptive and socio-ecological factors underlying the shift. This interpretation (or more accurately, speculation) suggests that modern apes also show reduced dimorphism compared with some Tertiary forms. The use of gorilla or orang measures as yardsticks to assess the fossils will then tend to underestimate within-species variation and so overestimate interspecies diversity.

There seem to be two approaches to resolving this. First, more fossil evidence is needed to make better estimates of ranges of variation and to identify species through *securely based* autapomorphies, i.e. characters that are not confined to just one or two specimens or are minor traits within some larger complex (see below). The other approach involves detailed studies of modern primates to investigate the range and *pattern* of dimorphism (e.g. Wood *et al.*, 1991) in order to get a better 'handle' on the fossils. This should allow identification of appropriate comparators, if such exist; recognition that they do not, and the extent of the mismatch, would itself be an advance.

Robusts, long thought to be a late evolving group, are now shown by KNM WT 17 000 to have differentiated at least by 2.6 my ago. This massive specimen, with its enlarged cheek teeth and unreduced anterior dentition, casts doubt (as does Stw 252) on the posited link between *A. afarensis* and *Homo* earlier argued on dental proportions (Johanson and White, 1979) but now abandoned. If the logic holds then these two specimens must also be involved in human ancestry, which would require a complete reappraisal of hominid systematics. Alternatively, the claimed dental apomorphies of *A. afarensis* and *Homo* are, in fact, plesiomorphies widely distributed among early hominids and therefore worthless as phyletic indicators.

There has recently been an increased tendency to distinguish between *A. robustus* (Kromdraai) and *A. crassidens* (Swartkrans) but if the former is older than, and ancestral to, the latter, then the pattern of dental/masticatory evolution in South African robusts was very different from that in East Africa. As with so much else more evidence is needed, but if confirmed by further discovery it strengthens the case for viewing the robusts in South and East Africa as distinct, parallel developments in response to similar selection pressures, rather than as conspecifics or very closely related (sibling) species.

Lomekwi and Kromdraai apart, most robust fossils date from around the Plio-Pleistocene boundary as, of course, do early *Homo* specimens, making it a period of unparalleled morphological diversity. There is growing awareness of the similarities between robust australopithecines and early *Homo* and while some (perhaps most) of these probably represent convergent adaptations, others suggest a close involvement in the origin of

Homo, perhaps a period of common ancestry post *A. afarensis/africanus*, which again takes us back to the early differentiation of robusts, and the phyletic status of KNM WT 17 000.

The outcome of this phyletic tangle is now becoming apparent in the final Pliocene record. It is increasingly clear that the early *Homo* sample is itself diverse with a need to recognise at least two species—*H. habilis* and *H. rudolfensis* (Wood, 1991)—quite apart from *H. erectus*. If the earliest representations of that grade are thought sufficiently distinct from the Asian and later African forms to warrant specific separation, then yet another species, *H. ergaster*, can be defined. Taxonomic issues aside, its presence at Turkana by *c.* 1.8 mya with a morphology that is difficult to derive from other known morphs without extensive homoplasy and within the available time-scale, provides further evidence of an array of early *Homo* species. Nor are contrasts confined to cranial/dental structures; the OH 62 and Nariokotome postcrania represent contrasting forms of body size, proportions and (presumably) locomotor behaviour. All these findings point to the conclusion that *Homo* as usually understood is polyphyletic, with the threshold crossed not just once but several times.

Recognition of multiple morphs prompts further questions about systematic status and phyletic relationships, adaptive responses and ecological interactions. As with earlier hominids, the fewer species posited, the greater the scale of required intraspecific variation, exceeding that of all modern forms, even the gorilla. If more fossil species need to be recognised, intraspecific range contracts but phyletic relationships become correspondingly more complex. Again, detailed cladistics as well as more fossils should help to identify groups better, but current studies give conflicting interpretations, partly because of differential character representation in the fossil record and partly because of necessary assumptions about polarities. Whatever view finally emerges, the old two-lineage model has, it seems, gone for good.

Multiple species also imply differing microevolutionary processes compared with the earlier model. Instead of one or two wide-ranging, polytypic species, several localised ones suggest a much greater role for local selection pressures with ecological or physical barriers impeding gene flow, and random factors having significant influence on the origin and extinction of individual populations. Reconstruction of the physical environments of early hominids coupled with community ecology approaches to faunal evolution offer promise for the reconstruction of selection pressures, as do primate socio-ecology studies that seek to establish associations between morphology, behaviour and ecology. However, despite claims to the contrary, e.g. Dunbar (1989) and Foley and Dunbar (1989), such approaches offer no golden key to unlock the truth; even with an extensive socio-ecological database the fossil groups and their morphological parameters need to be accurately defined before extrapolation to the inferred ecological correlates can be made. Few early hominid species imply generalist species with broad niches and associated behaviours, whereas a larger number suggest ecological specialists exploiting narrower niches and with contrasting socio-ecologies.

Above all, the increasingly convincing evidence of a radiating array of late Tertiary/basal Quaternary hominids rather than a single phyletic stem requires a shift of conceptual framework. Extreme anagenetic interpretations, most obviously represented by the 'single species hypothesis' with its underlying thread of a directed goal for hominid phylogeny, need to be discarded. Contrary to such formulations there was nothing inevitable about the evolution of *Homo sapiens*, and early hominids' initial cultural capabilities are unlikely to have conferred the adaptability and selective advantage that they subsequently afforded, so allowing a significant role to stochastic factors (see above). Viewed against the backdrop of mid-Miocene hominoid diversity, hominid evolution appears a restricted phenomenon, unimpressive except for the recent expansion of the extant species. From the later Miocene until less than 1 mya it is cercopithecoids, not hominids, and certainly not pongids, that are impressive in numbers, range and variety. On this perspective the low point of hominid evolution would be the Lower Pleistocene (*c.* 1.5–1.2 mya) with reduced species diversity, and hominids still confined to the savannah environments of South and East Africa. Just before this, *H. erectus/H. ergaster* appears; its origins are obscure, but it suggests a significant phyletic and adaptive shift within Hominidae. Larger bodied overall than many other early hominids, powerfully muscled (to judge by the Nariokotome skeleton and Olduvai Bed IV fossils) with distinctive cranio-facial morphology, its persistence presumably reflects a highly successful set of adaptations. The disappearance of other hominids shortly afterwards may be largely due to interhominid competition, or may simply be one aspect of the wider faunal turnover and reduced diversity apparent around this time.

Early hominid ecology was certainly complex, and is still poorly understood despite increasing information from detailed fieldwork. One factor that needs to be taken into account is the clear archaeological evidence of tool making by at least some early hominid groups, although whether this was species specific is unknown. One of the main achievements of the last decade has been the improved resolution that has followed from disentangling hominid activity from the results of taphonomic processes, so showing reconstructions of early hominids as proficient hunters to be wide of the mark: scavenging and only very limited carcass exploitation is a much more realistic reconstruction on available evidence. Environmental and archaeological data are still too few to tie hominid species to particular niches and so aid interpretation of skeletal morphology in relation to particular selection pressures.

One major indicator of growing adaptive success is hominid expansion beyond Africa. It is not clear what prompted such spread and whether climatic/environmental fluctuations were significant influences, although this seems likely. Given that Asian *H. erectus* fossils date between 1.0 and 0.25 mya, i.e. considerably later than the earlier African specimens, there is no need to posit a rapid, explosive radiation, but the process must certainly have involved some expansion of population numbers with increased intraspecific diversity as a result. In addition, wide geographical range, novel environments and associated selection pressures must have promoted more marked between-group differences. In particular, the contrasting environments (and possible contrasting adaptations, e.g. in tool kits) of South-East Asia (Pope, 1988) would certainly have resulted in at least intraspecific polytypism, but some have argued for specific difference, citing greater robustness and other features as distinctive apomorphies of Asian *H. erectus*. However, such claims need to be set against the intra-regional diversity of the specimens and the overall resemblance of the two continental groups.

Species recognition is also an issue when assessing variability in later Middle Pleistocene hominids. This is not just a question of the nature of the *H. erectus/H. sapiens* distinction but also whether further species recognition is warranted within what is generally called 'archaic *H. sapiens*' (Tattersall, 1986). This blanket term impedes appreciation of the diversity present within the material, so obscuring rather than clarifying groups and relationships. However, the nature and composition of any further groupings and the issue of whether specific distinction is warranted, remain to be demonstrated.

In fact consideration of Lower–Middle Pleistocene hominid diversity illustrates very well the influence of theoretical perspectives on interpretation. Cladistic techniques cannot accommodate within-group variation, and with small samples claimed apomorphies may be trivial and reflect nothing more than individual variability or local polymorphism, but they will nonetheless generate 'species' as an automatic consequence of the method. The claimed autapomorphies of Asian *H. erectus* and so its distinction from the early African material (Andrews, 1984; Wood 1984) appear to be a case in point (see chapter 7). A broadly phenetic approach can accommodate intra-group variation and change over time, but its use also has pitfalls. The bulk of the resemblance in such an analysis may reflect plesiomorphies rather than derived characters and group by primitive commonalities, so underestimating lineage diversity. The undifferentiated 'archaic *H. sapiens*' category above *may* be an example of this, as 'broad' *Homo habilis* appears to be.

In both cases, the best safeguard against errors is probably careful character selection to distinguish between primitive, derived and convergent traits together with evaluation of the scale and degree of within- and between-group variation to distinguish specific, polytypic and individual components contributing to the observed variation. It is also necessary to be aware of the likelihood of functional and/or logical correlation to avoid treating traits that repeatedly describe the same thing as being separate. Sample composition is another important consideration: excessive lumping as above results in over-coarse groupings that will fail to detect significant differences because the units of analysis are themselves inappropriate. This is an extreme case of the more general principle that intra- and inter-group characteristics naturally depend, above all else, upon group composition, which in turn requires some guiding principle for aggregation that may often be non-morphological (e.g. site location). Since systematic and evolutionary relationships are complex, no single method of phyletic inference can do justice to that complexity, and both cladistic and phenetic approaches are complementary in revealing different aspects of the evolutionary process (Bilsborough and Wood, 1986).

What of the pattern and velocity of hominid evolution ? Rate estimates naturally depend, above all else, on the phylogeny adopted. In fact, both punctuationists (e.g. Gould and Eldredge, 1977; Stanley, 1981; Eldredge and Tattersall, 1982) and gradualists (e.g. Cronin *et al.*, 1981) have claimed hominids as particularly clear examples of these two evolutionary models. However, the fact that the same specimens are cited to support both cases indicates that they are less than wholly convincing, and at present hominids cannot provide a test case for either—the record is still too meagre and the material too contentious. With small samples, in some cases individual specimens, and uncertainties about dating, the possibility of error is significant, and any patterns are at best suggestive, not definitive. In seeking to identify these it is important to recognise that selection of characters, methods and units of analysis may all bias the outcome of any investigation (see above). In particular, cladistic approaches will generate discontinuities suggestive of a punctuational interpretation, while the use of taxa as units will suppress intra-taxic diversity and focus on differences, so reinforcing impressions of a stop–go evolutionary pattern.

Again, consideration of *H. erectus* is illuminating. Rightmire (1984; 1985) has argued for stasis, and concludes that '*Homo erectus* fossils from sites dispersed widely across the Old World are in fact quite similar morphologically' (Rightmire, 1985). However, this is unsurprising given that his analysis focuses on those features that characterise *H. erectus*. Others have argued that the evidence indicates change, and sampling across a wide range of characters shows significant temporal patterning (Wolpoff, 1986). There *may* have been an acceleration in the later Middle Pleistocene around the *H. erectus/sapiens* boundary (clear evidence is lacking) but such episodic pulses, even if confirmed, can be accommodated by both evolutionary models. What is more significant is the failure of cladistic techniques to identify uniquely derived characters within archaic *H. sapiens* or provide evidence of hiatus between *H. erectus* and *H. sapiens*, both to be expected on the punctuational model. Equally significantly, the widespread occurrence of intermediate forms and similar morphological trends in all major continental areas favours the gradualist interpretation: in other words *H. erectus* and *H. sapiens* are chronospecies with an arbitrary interface reflecting anagenesis, not distinct clades. Periods of stasis are not clearly evident in the hominid fossil record.

Of course, there is not one evolution rate but many, for different characters evolve at varying velocities. Such mosaicism is an important aspect of detailed functional studies and also has the potential, especially with abundant environmental data, for insights into directional and stabilising selection intensities. However, studies of mosaicism have not fulfilled their early promise, largely because of uncertainties over phylogeny. Once the detailed pattern of human evolution is clearer, mosaic studies should aid reconstruction of selection pressures and adaptive responses to these.

The expanded fossil record of the last quarter of a million years reveals greater regional diversity, with distinct morphs in Europe, sub-Saharan Africa and Asia, although the first of these is by far the best known. The evolution of broadly modern morphology continues to be a focus of interest, and while the first appearance of such forms has been pushed back to *c*. 0.1 mya and more, the factors underlying the initial evolution of modern morphology and later disappearance of archaic forms remain uncertain. Upper Pleistocene evolutionary relationships are on a different, finer level than those of preceding periods, require correspondingly more complex interpretations, and provide more convincing evidence than hitherto of the inextricability of behavioural and morphological shifts in human evolution.

Molecular data, including the distribution of classical serological traits (e.g. Rh antigens), nuclear and some mt DNA studies, as well as linguistic data (Cavalli-Sforza *et al.*, 1988; Bowcock *et al.*, 1991; Cavalli-Sforza, 1991) all point to an African origin for modern humans (Stringer, 1990) as, indeed, does the fossil evidence (chapter 8). However, the timing and pattern of the subsequent spread of modern forms beyond Africa, and the processes underlying this, remain as controversial as ever. Most genetic analyses *posit* a predominant or exclusive role to drift, associated with constant evolution rates and full replacement, to account for observed genetic diversity. The fossil record is more suggestive of continuity (i.e. anagenesis with gene flow, see chapter 8) and molecular evolution could well be decoupled from morphological change, so that the former records the earlier expansion of *H. erectus*/archaic *H. sapiens* out of Africa, not the spread of modern humans (Van Valen, 1986).

Additional support for this view is provided by Eckhardt (1991) who points out discrepancies between morphological and mt DNA diversity in other groups (cichlid fishes). If mt DNA evolution is constant, as asserted by Cann *et al.* (1987), the data indicate that the differentiation of modern human populations must have been $>0.25-0.75$ mya (i.e. the *H. erectus* phase). Alternatively, mt DNA evolution is not constant, in which case it cannot be used to calibrate the timing and sequence of modern human origins. Even if constancy obtains the rate estimate based upon chimpanzee–hominid, likely divergence at 4–6 mya, which may extend to >7 mya, is obviously coarse for calibrating a recent event such as the origin of modern humans. In fact, arguments for the African origin of *H. s. sapiens* based on mt DNA data have been undermined by criticisms of the phyletic trees produced. Maddison (1991) showed there to be at least 10 000 trees with five fewer steps than Cann *et al.*'s (1987) shortest tree, while Templeton (1992) and Hedges *et al.* (1992) derive trees, including some of non-African origin, shorter than Vigilant *et al.*'s (1991) shortest tree. The earlier analyses allowed insufficient computer time to identify securely the shortest tree (Barinaga, 1992). The new, shorter trees differ appreciably from each other and the branching sequence cannot be resolved because of the nature of the data set (Hedges *et al.*, 1992).

If the pattern suggested by other genetic data is accurate, a similar patterning might be expected in at least some contemporary morphological traits, but none has been demonstrated. Further testing of competing models against modern data, both genetic and morphological, as well as an expanding fossil record should help resolve these uncertainties over modern human origins. Meanwhile mt DNA data are much more problematic sources for the time and place of modern human origins than appeared to be the case only a short time ago.

A more general development has been the recent move towards increased species recognition. This is most pronounced among Pliocene/basal Pleistocene hominids; but the tendency is also apparent in some analyses of later human evolution. Examples include the separation of African and Asian *H. erectus* grade fossils; Tattersall's (1986) arguments for splitting 'archaic' *H. sapiens*, and Stringer's proposal to distinguish Neanderthals as a separate species and restrict the term *H. sapiens* to anatomically modern forms. Others (e.g. Foley, 1989) have argued on general ecological grounds for multiple species in the hominid fossil record. The trend appears to be in part a reaction against previous extreme lumping schemes, partly an outcome of cladistic methods and partly because of increased awareness of contemporary species diversity, and the need to discriminate sample units as finely as possible. Some of these species appear to be proposed as 'morphospecies' reflecting discrete morphologies without any necessary genetic and evolutionary implications; others seem to be used in the accepted sense of 'biospecies', i.e. reproductively isolated genetic systems.

This approach has certainly not gone unopposed, and others (e.g. Rightmire, 1985; Smith, 1989; Wolpoff, 1989) have argued for a notably more parsimonious interpretation. Over the next few years a compromise might be achieved, with renewed recognition of distinct morphs among Middle and Upper Pleistocene hominids, but arguments for full specific distinctiveness (in the 'biospecies' sense) appear frankly unconvincing. Groves (1989), writing from a perspective informed by extensive systematic study of other groups (primate and non-primate), sees early hominid evolution as highly speciose, but his later (Mid–Upper Pleistocene) morphs are distinguished only as subspecies, not full species.

There have been major advances over the last two decades in the acquisition and analysis of evidence for human evolution and in the conceptual frameworks for handling the resulting information. The picture is both broader and more detailed, but has certainly not been simplified; it is, on the contrary, more complex. This is not retrogressive; while new discoveries naturally throw up new questions, and more of them, these are more sharply focused than before. Both the expanded evidence and increased uncertainties reflect the pace of change within palaeoanthropology: the search continues.

REFERENCES

Aeillo, L. and Dean, C. (1990) *An Introduction to Human Evolutionary Anatomy.* Academic Press, London and San Diego.

Aitken, M.J. (1985) *Thermoluminescence Dating.* Academic Press, London.

Aitken, M.J. (1990) *Science-based Dating in Archaeology.* Longman, London.

Andrews, P.J. (1984) An alternative interpretation of characters used to define *Homo erectus. Coursch. Forschungs. Senckenberg* **69**, 167–175.

Andrews, P. (1986a) Molecular evidence for catarrhine evolution. In: *Major topics in Primate and Human Evolution*, Wood, B., Martin, L. and Andrews, P. (eds.), Cambridge University Press, pp. 107–129.

Andrews, P. (1986b) Fossil evidence on human origins and dispersal. *Cold Spring Harbor Symp. Quant. Biol.* **51**, 419–428.

Andrews, P. (1990) Lining up the ancestors. *Nature* **345**, 664–665.

Andrews, P. and Cronin, J.E. (1982) The relationships of *Sivapithecus* and *Ramapithecus* and the evolution of the orang utan. *Nature* **297**, 541–546.

Andrews, P.J. and Tekkaya, J. (1980) A revision of the Turkish Miocene hominoid *Sivapithecus meteai. Palaeontology* **23**, 85–95.

Arambourg, C. and Coppens, Y. (1967) Sur le découverte dans le Pleistocene inférieur de la vallée de l'Omo (Ethiopie) d'une mandibule d'Australopithecien. *C.R. Acad. Sci.* **265-D**, 589–590.

Arensburg, B. (1989) New skeletal evidence concerning the anatomy of Middle Palaeolithic populations in the Middle East: the Kebara skeleton. In: *The Human Revolution: behavioral and biological perspectives on the origins of modern humans*, Mellars, P. and Stringer, C. (eds.), Edinburgh University Press, pp. 165–171.

Arensburg, B., Tilier, A.M., Vandermeersch, B., Duday, H. and Rak, Y. (1989) A Middle Palaeolithic human hyoid bone. *Nature* **338**, 758–761.

Aronson, J.L., Schmitt, T.J., Walter, R.C., Taieb, M., Tiercelin, J.J., Johanson, D.C., Naeser, C.W. and Nairn, A. (1977) New geochronologic and palaeomagnetic data for the hominid-bearing Hadar formation, Ethiopia. *Nature* **267**, 323–327.

Asfaw, B. (1983) A new hominid parietal from Bodo, Middle Awash Valley, Ethiopia. *Am. J. Phys. Anthrop.* **61**, 367–371.

Asfaw, B. (1985) Proximal femur articulation in Pliocene hominids. *Am. J. Phys. Anthrop.* **68**, 535–538.

Asfaw, B. (1987) The Belohdelie frontal: new evidence of early hominid cranial morphology from the Afar of Ethiopia. *J. Hum. Evol.* **16**, 611–624.

Ashlock, P.D. (1971) Monophyly and associated terms. *Syst. Zool.* **20**, 63–69.

Ashlock, P.D. (1979) An evolutionary systematist's view of classification. *Syst. Zool.* **28**, 441–450.

Barbetti, M. and Allen, H. (1972) Prehistoric Man at Lake Mungo, Australia by 32 000 BP. *Nature* **240**, 46–48.

Behrensmeyer, K. (1978) The habitat of Plio-Pleistocene hominids in East Africa: taphonomic and micro-stratigraphic evidence. In: *Early Hominids of Africa*, Jolly, C., (ed.), Duckworth, London, pp. 165–189.

Behrensmeyer, K. (1985) Taphonomy and palaeoecological reconstruction of hominid habitats in the Koobi Fora Formation. In: L'environment des hominides au Plio-Pleistocene. Masson, Paris, pp. 309–323.

Beynon, D. and Dean, M.C. (1988) Distinct dental development patterns in early fossil hominids. *Nature* **335**, 509–514.

Bilsborough, A. (1976) Patterns of evolution in Middle Pleistocene hominids. *J. Hum. Evol.* **5**, 423–439.

Bilsborough, A. (1978) Some aspects of mosaic evolution in hominids. In: *Recent Advances in Primatology. 3 Evolution*, Chivers, D.J. and Joysey, K.A. (eds.), Academic Press, London, pp. 335–350.

Bilsborough, A. (1983) The pattern of evolution within the genus *Homo*. In: *Progress in Anatomy*, Navaratnam, V. and Harrison, R.J. (eds.), Cambridge University Press, **3**, 143–164.

Bilsborough, A. (1986) Diversity, evolution and adaptation in early hominids. In: *Stone Age Prehistory*, Bailey, G.N. and Callow, P. (eds.), Cambridge University Press, pp. 197–220.

Bilsborough, A. and Wood, B.A. (1986) The nature, origin and fate of *Homo erectus*. In: *Major Topics in Primate and Human Evolution*, Wood, B.A., Martin, L. and Andrews, P. (eds.), Cambridge University Press, pp. 295–316.

Bilsborough, A. and Wood, B.A. (1988) Cranial morphometry of early hominids: facial regions. *Am. J. Phys. Anthrop.* **76**, 61–86.

Binford, L.R. (1973) Inter assemblage variability—the Mousterian and the 'functional' argument. In: *The Explanation of Culture Change in Prehistory*, Renfrew, C. (ed.), Duckworth, London, pp. 227–254.

Binford, L.R. (1981) *Bones: Ancient Men and Modern Myths*. Academic Press, New York.

Binford, L.R. (1984) *Faunal Remains From Klasies River Mouth*. Academic Press, Orlando.

Binford, S.L. and Binford, L.R. (1969) Stone tools and human behaviour. *Sci. Am.* **220(4)**, 70–84.

Binford, L. and Ho, C.K. (1985) Taphonomy at a distance: Zhoukoudien, 'the cave home of Beijing Man'? *Curr. Anthrop.* **26**, 413–442.

Binford, L. and Stone, N.M. (1986) Zhoukoudien: A closer look. *Curr. Anthrop.* **27**, 453–475.

Bischoff, J.L., Julia, R. and Mora, R. (1988) Uranium-series dating of the Mousterian occupation at Abric Romani, Spain. *Nature* **332**, 68–70.

Boaz, N.T. (1988) Status of *Australopithecus afarensis*. *Yearbook Phys. Anthrop.* **37**, 85–114.

Boaz, N.T. and Hampel, J. (1978) Strontium content of fossil tooth enamel and diet of early hominids. *J. Paleontol.* **52**, 928–933.

Boaz, N.T. and Howell, F.C. (1977) A gracile hominid cranium from Upper Member G of the Shungura Formation, Ethiopia. *Am. J. Phys. Anthrop.* **46**, 93–108.

Bonde, N. (1977) Cladistic classification as applied to vertebrates. In: *Major Patterns in Vertebrate Evolution*, Hecht, M.K., Goody, P.C. and Hecht, B.M. (eds.), Plenum, New York, pp. 741–804.

Bonis, L. de, Bouvrain, G., Geraads, D. and Koufos, G. (1990) New hominid skull material from the late Miocene of Macedonia in Northern Greece. *Nature* **345**, 712–714.

Bordes, F. (1961) Mousterian cultures in France. *Science.* **134**, 803–810.

Bordes, F. (1968) *The Old Stone Age*. Weidenfeld, London.

Boule, M. (1911–13) L'homme fossile de la Chapelle-aux-Saints *Annales de Paleontologie* **6**, 11–172; **7**, 21–192; **8**, 1–70.

Boule, M. and Vallois, H.V. (1957) *Fossil Men* (4th edn). Thames and Hudson, London.

Bowcock, A.M., Kidd, J.R., Mountain, J.L., Hebert, J.M., Carotenuto, L., Kidd, K.K. and Cavalli-Sforza, L.L. (1991) Drift, admixture, and selection in human evolution: a study with DNA polymorphisms. *Proc. Natl Acad. Sci. USA* **88**, 839–843.

Bowen, B.E. and Vondra, C.F. (1973) Stratigraphical relationships of the Plio-Pleistocene deposits, East Rudolf, Kenya. *Nature* **242**, 391–393.

Bowler, J.M., Thorne, A.G. and Pollach, H.A. (1972) Pleistocene man in Australia: age and significance of the Mungo skeleton. *Nature* **240**, 48–50.

Brace, C.L. (1963) Structural reduction in evolution. *Am. Naturalist* **97**, 39–49.

Brace, C.L., Shao Xiang-ging and Zhang Zhen-biao (1984) Prehistoric and modern tooth size in China. In: *The origins of modern humans: a world survey of the fossil evidence*, Smith, F.H. and Spencer, F. (eds.), A.R. Liss, New York, pp. 485–516.

Brain, C.K. (1970) New fields at the Swartkrans site. *Nature* **225**, 1112–1119.

Brain, C.K. (1981) *The hunters or the hunted? An introduction to African cave taphonomy.* Chicago University Press.

Brain, C.K. (1988) New information from the Swartkrans Cave of relevance to 'robust' australopithecines. In: *Evolutionary History of the 'robust' Australopithecines*, Grine, F.E (ed.), Aldine de Gruyter, New York, pp. 317–316.

Brauer, G. (1984) A craniological approach to the origin of anatomically modern *homo sapiens* in Africa and implications for the appearance of modern Europeans. In: *The origin of modern humans: a world survey of the fossil evidence.* Smith, F.H. and Spencer, F. (eds.), A.R. Liss, New York, pp. 387–410.

Brauer, G. (1989) The evolution of modern humans: a comparison of the African and non-African evidence. In: *The human revolution: behavioural and biological perspectives on the origins of modern humans*, Mellars, P. and Stringer, C. (eds.), Edinburgh University Press, pp. 123–154.

Bromage, T.D. and Dean, M.C. (1985) Re-evaluation of the age of death of Plio-Pleistocene hominids. *Nature* **317**, 525–528.

Broom, R. (1937) The Sterkfontein ape. *Nature* **139**, 326.

Broom, R. (1938) The Pleistocene anthropoid apes of South Africa. *Nature* **142**, 377–379.

Broom, R. (1949) Another new type of fossil ape man. *Nature* **163**, 57.

Broom, R. and Robinson, J.T. (1952) Swartkrans ape-man, *Paranthropus crassidens. Transv. Mus. Men* **6**, 1–124.

Broom, R. and Schepers, G.W.H. (1946) The South African fossil ape men, the Australopithecinae. *Transv. Mus. Mem.* **2**, 1–272.

Brown, P. (1981) Artificial cranial deformation: a component in the variation in Pleistocene Australian aboriginal crania. *Archaeology in Oceania* **16**, 156–167.

Brown, F.H. (1982) Tulu Bor tuff at Koobi Fora correlated with the Sidi Harkoma tuff at Hadar. *Nature* **300**, 631–632.

Brown, F.H. and Feibel, C.S. (1986) Revision of lithostratigraphic nomenclature in the Koobi Fora region, Kenya. *J. Geol. Soc.* **143**, 297–310.

Burr, D.B. (1976) Neanderthal vocal tract reconstruction. A critical appraisal. *J. Hum. Evol.* **5**, 285–290.

Butzer, K. (1974) Paleoecology of South African australopithecines: Taung revisited. *Curr. Anthrop.* **15**, 367–388.

Byrne, R. and Whiten, A. (eds) (1988) Machiavellian intelligence: social expertise and the evolution of intellect in monkeys, apes and humans. Clarendon Press, Oxford.

Campbell, B.G. (1966) *Human Eolution.* Heinemann, London.

Cann, R.L. (1988) DNA and human origins. *Ann. Rev. Anthrop.* **17**, 127–143.

Cann, R.L., Stoneking, M. and Wilson, A.C. (1987) Mitochondrial DNA and human evolution. *Nature* **325**, 31–36.

Cavalli-Sforza, L.L. (1991) Genes, peoples and languages. *Sci. Am.* **265(5)**, 72–79.

Cavalli-Sforza, L.L., Piazza, A., Menozzi, P. and Mountain, J. (1988) Reconstruction of human evolution: bringing together genetic, archaeological, and linguistic data. *Proc. Natl Acad. Sci. USA* **85**, 6002–6006.

Chamberlain, A.T. (1987) *A taxonomic review and phylogentic analysis of Homo habilis.* Ph.D. thesis, University of Liverpool.

Chamberlain, A.T. and Wood, B.A. (1987) Early hominid phylogeny. *J. Hum. Evol.* **16**, 119–133.

Chase, P.G. and Dibble, H.L. (1987) Middle Paleolithic symbolism: a review of current evidence and interpretations. *J. Anthropol. Archaeol.* **6**, 263–296.

Chavaillon, J., Chavaillon, N., Hours, F. and Piperno, M. (1979) From the Olduwan to the Middle Stone Age at Melka Kunture (Ethiopia): Understanding cultural changes. *Quaternaria* **21**, 87–114.

Clark, J.D. (1987) Transitions: *Homo erectus* and the Acheulian: the Ethiopian sites of Gadeb and the Middle Awash. *J. Hum. Evol.* **16**, 809–826.

Clark, J.D. and Kurashina, H. (1979) Hominid occupation of the east central highlands of Ethiopia in the Plio-Pleistocene. *Nature* **282**, 33–39.

Clarke, R.J. (1978) A new *Australopithecus* cranium from Sterkfontein and its bearing on the ancestry of *Paranthropus*. In: *Evolutionary History of the 'robust' Australopithecines*, Grine, F.E. (ed.), Aldine de Gruyter, New York.

Clarke, R.J. and Howell, F.C. (1972) Affinities of the Swartkrans hominid cranium. *Am. J. Phys. Anthrop.* **37**, 319–336.

Clausen, I.H.S. (1989) Cranial capacity in *Homo erectus*: stasis or non-stasis, direction of trend. In: *Hominidae: Proc. 2nd Int. Cong. Human Paleontology*, Giacobini, G. (ed.), Jaca, Milan, pp. 217–220.

Conroy, G.C. (1990) *Primate Evolution.* W.W. Norton & Co., New York.

Coon, C.S. (1962) *The Origin of Races.* A. Knopf, New York.

Corruccini, R.S. and Ciochan, R.L. (1979) Primate facial allometry and interpretations of australopithecine variation. *Nature* **281**, 62–64.

Corruccini, R.S. and McHenry, H.M. (1980) Cladometric analysis of Pliocene hominids *J. Hum. Evol.* **9**, 209–221.

Cronin, J.E. (1983) Apes, humans and molecular clocks: a reappraisal. In: *New Interpretations of Ape and Humans Ancestry*, Ciochon, R.L. and Corruccini, R.S. (eds.), Plenum, New York, pp. 115–136.

Cronin, J.E, Boaz, N.T., Stringer, C.B. and Rak, Y. (1981) Tempo and mode in hominid evolution. *Nature* **292**, 113–122.

Dart, R.A. (1925) *Australopithecus africanus*: The man-ape of South Africa. *Nature* **115**, 195–199.

Day, M.H. (1971) Postcranial remains of *Homo erectus* from Bed IV Olduvai Gorge, Tanzania. *Nature* **221**, 230–233.

Day, M.H. (1985) Pliocene hominids. In: *Ancestors: the Hard Evidence*, Delson, E. (ed.), A.R. Liss, New York, pp. 91–93.

Day, M.H. and Stringer, C.B. (1982) A reconsideration of the Omo Kibish remains and the *erectus-sapiens* transition. In: *L' Homo erectus et la place de l'homme de Tauteval parmi les hominides fossils,* de Lumley, M.A. (ed.), CNRS, Nice, pp. 814–846.

Dean, M.C. (1987) The growth layers and incremental markings in hard tissues: a review of the literature and some preliminary observations about enamel structure in *Paranthropus boisei*. *J. Hum. Evol.* **16**, 157–172.

Delson, E. (1985) Palaeobiology and age of African *Homo erectus*. *Nature* **316**, 762–763.

Delson, E. (1988) Chronology of South Africa Australopith site. In: *Evolutionary History of the 'Robust' Australopithecines*, Grine, F.E. (ed.), Aldine de Gruyter, New York, pp. 317–324.

Delson, E., Eldredge, N. and Tattersall, I. (1977) Reconstruction of hominid phylogeny: a testable framework based on cladistic analysis. *J. Hum. Evol.* **6**, 263–278.

Demes, B. (1987) Another look at an old face: biomechanics of the neanderthal facial skeleton reconsidered. *J. Hum. Evol.* **16**, 297–303.

De Villiers, H. and Fatti, L.P. (1982) The antiquity of the negro. *S. Afr. J. Sci.* **78**, 212–215.

Dobzhansky, T. (1962) *Mankind Evolving: the Evolution of the Human Species.* Yale University Press.

Dunbar, R.I.M. (1989) Ecological modelling in an evolutionary context. *Folia Primatol.* **53**, 235–246.

Eckhardt, R.B. (1991) Fish scales for human origins. *Nature* **349**, 112.

Eisenberg, J.F. (1981) *The Mammalian Radiations: Analysis of Trends in Evolution, Adaptation and Behaviour*. Athlone, London.

Eldredge, N. and Gould, S.J. (1972) Speciation and punctuated equilibria: an alternative to phyletic gradualism. In: *Models in Paleobiology*, Schopf, T.J. (ed.), W.H. Freeman, San Francisco, pp. 82–115.

Eldredge, N. and Tattersall, I. (1982) *The Myths of Human Evolution*. Columbia University Press, New York.

Falk, D. (1983) Cerebral Cortices of East African early hominids. *Science* **221**, 1072–1074.

Falk, D. (1985) Hadar AL 162-28 endocast as evidence that brain enlargement preceded cortical reorganisation in hominid evolution. *Nature* **313**, 45–47.

Falk, D. (1986) Endocast morphology of Hadar Hominid AL 162-28. *Nature* **321**, 536–537.

Feibel, C.S., Brown, F.H. and McDougall, I. (1989) Stratigraphic context of fossil hominids from the Omo Group deposits: Northern Turkana Basin, Kenya and Ethiopia. *Am. J. Phys. Anthrop.* **78**, 595–622.

Fleagle, J.G. (1988) *Primate Adaptation and Evolution*. Academic Press, San Diego.

Foley, R.A. (ed) (1984) *Hominid Evolution and Community Ecology*. Academic Press, London.

Foley, R.A. (1987) *Another Unique Species: Patterns in Human Evolutionary Ecology*, Longman, Harlow.

Foley, R. (1989) The ecological conditions of speciation: a comparative approach to the origins of anatomically-modern humans. In: *The Human Revolution: Behavioural and Biological Perspectives on the Origins of Modern Humans*, Mellars, P. and Stringer, C. (eds.), Edinburgh University Press, pp. 298–318.

Foley, R.A. and Dunbar R. (1989) Beyond the bones of contention. *New Scientist* **144**, 37–41.

Foley, R.A. and Lee, P. (1989) Finite social space, evolutionary pathways and reconstructing hominid behaviour. *Science* **243**, 901–906.

Frayer, D.W. (1977) Metric dental change in the European Upper Paleolithic and Mesolithic. *Am. J. Phys. Anthrop.* **46**, 109–120.

Frayer, D.W. (1984) Biological and cultural change in the European late Pleistocene and early Holocene. In: *The origins of modern humans: a world survey of the fossil evidence*. Smith, F.H. and Spencer, F. (eds.), A.R. Liss, New York, pp. 211–250.

Gamble, C. (1986) *The Palaeolithic settlement of Europe*. Cambridge University Press.

Garrod, D.A.E. and Bate, D. (1937) *The Stone Age of Mt Carmel* Vol. I Oxford University Press.

Geschwind, N. (1972) Language and the brain. *Sci. Am.* **226**, 76–83.

Gingerich, P. (1978) Phylogeny reconstruction and the phylogenetic position of Tarsiers. In: *Recent Advances in Primatology 3: Evolution*, Chivers, D.J. and Joysey, K.A. (eds.), Academic Press, London, pp. 249–256.

Gingerich, P. and Schoeninger, M. (1977) The fossil record and primate phylogeny *J. Hum. Evol.* **6**, 483–505.

Goodman, M., Baba, M.L. and Darga, L.L. (1983) The bearing of molecular data on the cladogenesis and times of divergence of hominoid lineages. In: *New Interpretations of Ape and Human Ancestry*, Ciochon, R.L. and Corruccini, R.S. (eds.), Plenum, New York, pp. 67–86.

Gould, S.J. and Eldredge, N. (1977) Punctuated equilibria: the tempo and mode of evolution reconsidered. *Paleobiology* **3**, 115–151.

Gowlett, J.A., Harris, J.H.K., Walton, D. and Wood, B.A. (1981) Early archaeological sites, hominid remains and traces of fire from Chesowanja, Kenya. *Nature* **294**, 125–129.

Grine, F.E. (1981) Trophic differences between 'gracile' and 'robust' australopithecines: a scanning electron microscope analysis of occlusal events. *S. Afr. J. Sci.* **77**, 203–230.

Grine, F.E. (1988a) Evolutionary history of the 'robust' australopithecines: a summary and historical perspective. In: Grine, *Evolutionary History of the 'Robust' Australopithecines*, Grine, F.E. (ed.), Aldine de Gruyter, New York, pp. 509–520.

Grine, F.E. (1988b) *Australopithecus*. In: *Encyclopedia of Human Evolution and Prehistory*, Tattersall, I., Delson, E. and van Couvering, J. (eds.), Garland, New York and London, pp. 67–74.

Groves, C.P. (1989) *A Theory of Human and Primate Evolution*. Clarendon Press, Oxford.

Groves, C.P. and Mazak, V. (1975) An approach to the taxonomy of the Hominidae: gracile villafranchian hominids to Africa. *Casopis pro Mineralogii Geologii* **20**, 225–247.

Grun, R. and Stringer, C.B. (1991) Electron spin resonance dating and the evolution of modern humans. *Archaeometry* **33**, 153–199.

Habgood, P. (1985) The origin of the Australian aborigines: an alternative approach and view. In: *Hominid Evolution past, present and future*, Tobias, P.V. (ed.), A.R. Liss, New York, pp. 367–380.

Habgood, P. (1989) The origin of anatomically modern humans in Australia. In: *The Human Revolution: behavioural and biological perspectives on the origins of modern humans*, Mellars, P. and Stringer, C. (eds.), Edinburgh University Press, pp. 245–273.

Harris, J.W.K., Williams, P.G., Verniers, J., Tappen, M.J., Stewart, K., Helgren, B., de Heinzelin, J., Boaz, N.T. and Bellomo, R. (1987) Late Pliocene hominid occupation in Central Africa: the setting, context and character of the Senga 5A site, Zaire. *J. Hum. Evol.* **16**, 701–728.

Hay, R.L. (1976) *Geology of the Olduvai Gorge*. University of California Press, Berkeley.

Hedges, R.E.M. and Gowlett, J.A.J. (1986) Radiocarbon dating by accelerator mass spectrometry. *Sci. Am.* **254**, 82–89.

Heim, J.L. (1976) *Less Hommes Fossiles de La Ferrassie, I*. Masson, Paris.

Hennig, W. (1966) *Phylogenetic Systematics*. University of Illinois Press, Chicago.

Hill, A. (1987) Causes of perceived faunal change in the later Neogene of East Africa. *J. Hum. Evol.* **16**, 583–596.

Hill, A. and Ward, S. (1988) Origin of the Hominidae: the record of African large hominid evolution between 14 my and 4 my. *Yearbook of Phys. Anthrop.* **31**, 49–84.

Holloway, R.L. (1972a) New australopithecine endocast, SK 1585, from Swartkrans, S. Africa. *Am. J. Phys. Anthrop.* **37**, 173–186.

Holloway, R.L. (1972b) Australopithecine endocasts, brain evolution in the Hominidae, and a model of hominid evolution. In: *Functional and Evolutionary Biology of Primates*, Tuttle, R. (ed.), Aldine, New York.

Holloway, R.L. (1974) The casts of fossil hominid brains. *Sci. Am.* **231**, 106–115.

Holloway, R.L. (1983) Cerebral brain endocast pattern of *Australopithecus afarensis* hominid. *Nature* **303**, 420–422.

Holloway, R.L. (1988) 'Robust' australopthecine brain endocasts: some preliminary observations. In: *Evolutionary History of the 'Robust' Australopithecines*, Grine, E.F. (ed.), Aldine de Gruyter, New York, pp. 97–106.

Holloway, R.L. and de la Coste Lareymondie, M.C. (1982) Brain endocast asymmetry in pongids and hominids: some preliminary findings on the paleontology of cerebral dominance. *Am. J. Phys. Anthrop.* **58**, 101–110.

Holloway, R.L. and Kimbel, W.H. (1986) Endocast morphology of Hadar Hominid AL 162–28. *Nature* **321, 536.**

Holloway, R.L. and Post, D.C. (1982) The relativity of relative measures and hominid mosaic evolution. In: *Primate Brains: Evolution, Methods and Concepts*, Armstrong, E. and Falk, D. (eds.), Plenum, New York, pp. 57–76.

Howell, F.C. (1952) Pleistocene glacial ecology and the evolution of 'classic Neanderthal' man. *Southwestern J. Anthropology* **8**, 377–410.

Howell, F.C. (1960) European and northwest African Middle Pleistocene hominids. *Curr. Anthrop.* **1**, 195–232.

Howell, F.C. (1978) Hominidae. In: *Evolution of African Mammals*, Maglio, V.J. and Cooke, H.B.S. (eds.), Harvard university Press, Cambridge, pp. 154–248.

Howell, F.C. (1981) Some views of *Homo erectus* with special reference to its occurrence in Europe. In: *Homo erectus: Papers in Honor of Davidson Black*. Sigmon, B.A. and Cybulski, S. (eds.), University of Toronto Press, pp. 153–157.

Howell, F.C., Haesaerts, P. and de Heinzelin, J. (1987) Depositional environments, archaeological occurrences and hominids from Members E and F of the Shunguru Formation (Omo basin, Ethiopia). *J. Hum. Evol.* **16**, 665–700.

Howells, W.W. (1973) *Cranial variation in man. A study by multivariate analysis of patterns of difference among recent human populations*. *Paper Peabody Museum, Harvard University*, Vol. 67.

Howells, W.W. (1976) Explaining modern man: evolutionists versus migrationists. *J. Hum. Evol.* **5**, 477–495.

Hublin, J.J. (1985) Human fossils from the North African Middle Pleistocene and the origin of *Homo sapiens*. In: *Ancestors: the Hard Evidence*. Delson, E. (ed.), A.R. Liss, New York, pp. 283–288.

Hull, D.B. (1979) The limits of cladism. *Syst. Zool.* **28**, 416–440.

Isaac, G.L. (1978) The food sharing behavior of protohuman hominids. *Sci. Am.* **238**, 90–108.

Isaac, G.L. (1981) Emergence of human behaviour patterns. *Phil. Trans Roy. Soc. B.* **292**, 177–188.

Isaac, G.L. and Crader, D. (1981) To what extent were early hominids carnivorous: an archaeological perspective. In: *Omnivorous Primates: gathering and hunting in human evolution*. Harding, R.S.O. and Teleki, G. (eds.), Columbia, New York, pp. 37–103.

Jacob, T. (1973) Morphology and paleoecology of early man in Java. *Int. Cong. Anthrop. Ethnol. Sci. 9*, Chicago.

Jacob, T. (1976) Early population in the Indonesian region. In: *The Origin of the Australians*, Kirk, R.L. and Throne, A.G. (eds.), Aust. Inst. Aboriginal Studies, Canberra, pp. 81–93.

Jeffreys, A.J. (1989) Molecular biology and human evolution. In: *Human Origins*, Durant, J.R. (ed.), Clarendon Press, Oxford, pp. 27–52.

Jelinek, J. (1978) *Homo erectus or Homo sapiens?* In: *Recent Advances in Primatology 3, Evolution*, Chivers, D.J. and Joysey, K.A. (eds.), Academic Press, London, pp. 419–430.

Jelinek, J. (1980) Variability and geography: contribution to our knowledge of European and North African Middle Pleistocene hominids. *Anthropologie (Brno)* **18**, 109–114.

Jelinek, J. (1985) Human evolution. In: *Hominid evolution past, present and future*, Tobias, P.V. (ed.), A.R. Liss, New York, pp. 341–354.

Jerison, H.J. (1973) *Evolution of the Brain and Intelligence*. Academic Press, London.

Johanson, D.C. and White, T.D. (1979) A systematic assessment of early African hominids. *Science* **202**, 321–330.

Johanson D.C., White, T.D. and Coppens, Y. (1978) A new species of the genus *Australopithecus* (Primates: Hominidae) from the Pliocene of eastern Africa. *Kirtlandia* **28**, 1–14.

Johanson, D.C., Taieb, M. and Coppens, Y. (1982a) Pliocene hominids from the Hadar formation, Ethiopia (1973–1977): Stratigraphic, chronologic, and paleoenvironmental contexts with notes on hominid morphology and systematics. *Am. J. Phys. Anthrop.* **57**, 373–402.

Johanson, D.C., Taieb, M., Coppens, Y., Lovejoy, C.O., Kimbel, W.H., White, T.D., Ward, S.C., Bush, M. and Latimer, B.M. (1982b) Pliocene Hominid Fossils from Hadar, Ethiopia. *Am. J. Phys. Anthrop.* **57**, 373–720.

Johanson, D.C., Masao, F.T., Eck, G.G., White, T.D., Walter, R.C., Kimbel, W.H., Asfaw, B., Manega, P., Ndessokia, P. and Suwa, G. (1987) New partial skeleton of *Homo habilis* from Olduvai Gorge, Tanzania. *Nature* **327**, 205–209.

Jolly, C.J. (1970) The seed eaters: a new model of hominid differentiation based on a baboon analogy. *Man* **5**, 5–26.

Jungers, W. (1988a) Relative joint size and hominoid locomotor adaptations with implications for the evolution of hominid pibedalism. *J. Hum. Evol.* **17**, 247–265.

Jungers, W.L. (1988b) New estimates of body size in australopithecines. In: *Evolutionary History of the 'Robust' Australopithecines*, Grine, F.E. (ed.), Aldine de Gruyter, New York, pp. 115–126.

Kalb, J.E., Jolly, C., Mebrate, A., Tebedge, S., Smart, C., Oswald, E.B., Cramer, D., Whitehead, P., Wood, C.B., Conroy, G.C., Adefris, T., Sperling, L. and Kana, B. (1982) Fossil mammals and artifacts from the Awash Group, Middle Awash Valley, Afar, Ethiopia. *Nature* **298**, 25–29.

Kay, R. and Grine, F.E. (1988) Tooth Morphology: wear and diet in *Australopithecines* and *Paranthropus* from Southern Africa. In: *Evolutionary History of the 'Robust' Australopithecines*, Grine, F.E. (ed.), Aldine de Gruyter, New York.

Kelley, J. and Xu, Q. (1991) Extreme sexual dimorphism in a Miocene hominoid. *Nature* **352**, 151–153.

King, M.C. and Wilson, A.C. (1975) Evolution at two levels in humans and chimpanzees. *Science* **188**, 107–116.

Klein, R.G. (1988) The causes of 'robust' Australopithecine extinction. In: *Evolutionary History of the 'Robust' Australopithecines*, Grine, F.E. (ed.), Aldine de Gruyter, New York, pp. 499–505.

Klein, R.G. (1989) *The Human Career: Human Biological and Cultural Origins*. Chicago University Press, Chicago.

Kimbel, W. and White, T.D. (1988) Variation, sexual dimorphism and the taxonomy of *Australopithecus*. In: *Evolutionary History of the 'Robust' Australopithecines*, Grine, F.E. (ed.), Aldine de Gruyter, New York, pp. 175–192.

Kimbel, W. H., White, T.D. and Johanson, D.C. (1984) Cranial morphology of *Australopithecus afarensis*: a comparative study based on a composite reconstruction of the adult skull. *Am. J. Phys. Anthrop.* **65**, 337–388.

Kimbel, W.H., White, T.D. and Johanson, D.C. (1985) Craniodental morphology of the hominids from Hadar and Laetoli: evidence of '*Paranthropus* 'and *Homo* in the Mid-Pliocene of Eastern Africa? In: *Ancestors: the Hard Evidence*, Delson, E. (ed.), A.R. Liss, New York, pp. 120–137.

Kimbel, W.H., White, T.D. and Johanson, D.C. (1988a) Implications of KNM WT 17 000 for the evolution of 'robust' *Australopithecus*. In: *Evolutionary History of the 'Robust' Australopithecines*, Grine, F.E. (ed.), Aldine de Gruyter, New York, pp. 259–268.

Kimbel, W.H., White, T.D. and Johanson, D.C. (1988b) *J. Hum. Evol.*

Koenigswald, G.H.R. von (1950) Fossil hominids from the Lower Pleistocene of Java. *Proc. Int. Geol. Cong. Sect. 9.* 59–61.

Koenigswald, G.H.R. von (1975) Early Man in Java: catalogue and problems. In: *Paleontology, Morphology and Paleoecology*, Tuttle, R. (ed.), Mouton, The Hague, pp. 303–310.

Leakey, L.S.B. (1960) Recent discoveries at Olduvai Gorge. *Nature* **188**, 1050–1052.

Leakey, L.S.B. (1961a) New finds at Olduvai Gorge. *Nature* **189**, 649–650.

Leakey, L.S.B. (1961b) The juvenile mandible from Olduvai. *Nature* **191**, 417–418.

Leakey, L.S.B. (1966) *Homo habilis, Homo erectus* and the australopithecines. *Nature* **209**, 1279–1281.

Leakey, M.D. (1971) *Olduvai Gorge Vol. 3 Excavations in Beds I and II 1960–1963*. Cambridge University Press, p. 267.

Leakey, R.E.F. (1976a) *Australopithecus, Homo erectus* and the single species hypothesis. *Nature* **261**, 572–576.

Leakey, R.E.F. (1976b) An overview of the Hominidae from East Rudolf, Kenya. In: *Earliest Man and Environments in the Lake Rudolf Basin*, Coppens, Y., Howell, F.C., Isaac, G.L. and Leakey, R.E.F. (eds.), Chicago University Press, pp. 476–483.

Leakey, R. and Leakey, M. (1986a) A new Miocene hominoid from Kenya. *Nature* **324**, 143–146.

Leakey, R. and Leakey, M. (1986b) A second new Miocene hominoid from Kenya. *Nature* **324**, 146–148.

Leakey, R.E.F. and Walker, A.C. (1985) Further hominids from the Plio-Pleistocene of Koobi Fora, Kenya. *Am. J. Phys. Anthrop.* **67**, 135–164.

Leakey, R.E.F. and Walker, A.C. (1989) *Early Homo erectus* from West Lake Turkana, Kenya. In: *Hominidae. Proc. 2nd Int. Cong. Human Paleontology*, Giacobini, G. (ed.), Jaca, Milan, pp. 209–216.

Leakey, L.S.B., Tobias, P.V. and Napier, J.R. (1964) A new species of the genus *Homo* from Olduvai Gorge. *Nature* **202**, 7–9.

Leakey, M.D., Tobias, P.V., Martyn, J.E. and Leakey, R.E. (1969) An Acheulian industry with prepared core technique and the discovery of a contemporary hominid at Lake Baringo, Kenya. *Proc. Prehist. Soc.* **25**, 48–76.

Leakey, R.E.F., Leakey, M.G. and Behrensmeyer, A.K. (1978) The hominid catalogue. In: *Koobi Fora Research Project* I: *The fossil hominids and an introduction to their context*, Leakey, M.G. and Leakey, R.E. (eds.), Clarendon Press, Oxford, pp. 86–182.

Leakey, R.E.F., Walker, A.C., Ward, C.V. and Grausz, H.M. (1989) A partial skeleton of a gracile hominid from the Upper Burgi Member of the Koobi Fora Formation, East Lake Turkana, Kenya. In: *Hominidae Proc. 2nd Int. Cong. of Human Paleontology*, Giacobini, G. (ed.), Jaca, Milan, pp. 167–173.

Lestrel, P.E. (1976) Hominid brain size versus time: revised regression estimates. *J. Hum. Evol.* **5**, 207–212.

Leutenegger, W. (1972) Newborn size and pelvic dimensions of *Australopithecus*. *Nature* **240**, 568–569.

Leutenegger, W. (1987) Neonatal brain size and neurocranial dimensions in Pliocene hominids: implications for obstetrics. *J. Hum. Evol.* **16**, 291–296.

Lévêque, F. and Vandermeersch, B. (1981) Le Neandertalien de Saint-Césaire. *Récherché* **12**, 242–244.

Lewin, R. (1981) Ethiopian stone tools are world's oldest. *Science* **211**, 806–807.

Lieberman, P. (1975) *On the origins of language: an introduction to the evolution of human speech.* Macmillan, New York.

Lieberman, P., Crelin, E.S. and Klatt, D.S. (1972) Phonetic ability and related anatomy of the newborn, adult human, Neanderthal man and the chimpanzee. *Am. Anthropologist* **74**, 287–307.

Lieberman, D.E., Pilbeam, D.R. and Wood, B.A. (1988) A probabilistic approach to the problem of sexual dimorphism in *Homo habilis*: a comparison of KNM ER 1470 and KNM ER 1813. *J. Hum. Evol.* **17**, 503–511.

Lovejoy, C.O. (1974) The gait of australopithecines. *Yearbook of Physical Anthropology* **17**, 147–161.

Lovejoy, C.O. (1981) The origin of man. *Science* **211**, 341–350.

Lovejoy, C.O. (1988) Evolution of ,human walking. *Sci. Am.* **259**, 82–89.

McCown, T.D. and Keith, K. (1939) *The Stone Age of Mt Carmel* Vol. 2 Clarendon, Oxford.

McGuire, B. (1980) Further observations on the nature and provenance of the lithic artefacts from the Makapansgat limeworks. *Palaeontologic Africana* **23**, 127–157.

McHenry, H.M. (1986) The first bipeds: a comparison of the *A. afarensis* and *A. africanus* postcranium and implication for the evolution of bipedalism. *J. Hum. Evol.* **15**, 177–191.

McHenry, H.M. (1988) New estimates of body weight in early hominids and their significance to encephalization and megodentia in 'robust' australopithecines. In: *Evolutionary History of the 'Robust' Australopithecines*, Grine, F.E. (ed.), Aldine de Gruyter, New York, pp. 133–148.

Mann, A.E. (1975) *Paleodemographic aspects of the South African Australopithecines.* University of Pennsylvania Publications in Anthropology.

Mann, A.E., Lampl, M. and Monge, J. (1990) Patterns of ontogeny in human evolution: evidence from dental development. *Yearbook of Phys. Anthrop.* **33**, 111–150.

Marks, A.E. (1983) The Middle to Upper Palaeolithic transition in the Levant. *Advances in World Archaeology* **2**, 51–98.

Martin, L. (1991) Teeth, sex and species. *Nature* **352**, 111–112.

Martin, R.D. (1990) *Primate origins and evolution: a phylogenetic reconstruction.* Princeton University Press, Princeton, N.J.

Mayr, E. (1981) Biological classification: towards a synthesis of opposing methodologies. *Science* **214**, 510–516.

Mellars, P.A. (1970) The chronology of Mousterian industries in the Perigord region of S.W. France. *Proc. Prehist. Soc.* **35**, 134–171.

Mellars, P.A. (1973) The character of the Middle Upper Palaeolithic transition in S.W. France. In: *The explanation of culture change in prehistory*, Renfrew, C. (ed.), Duckworth, London, pp. 255–276.

Mellars, P. (1986) A new chronology for the French Mousterian period. *Nature* **322**, 410–411.

Mellars, P. (1989) Major issues in the emergence of modern humans. *Curr. Anthrop.* **30**, 349–385.

Mercier, N., Valladas, H., Joron, J.-L., Reyss, J.-L., Lévêque, F. and Vandermeersch, B. (1991) Thermoluminescence dating of the late Neanderthal remains from St Césaire. *Nature* **351**, 737–739.

Miller, J.A. (1991) Does brain size variability provide evidence of multiple species in *Homo habilis*? *Am. J. Phys. Anthrop.* **84**, 385–398.

Napier, J.R. (1980) *Hands.* George Allen and Unwin., London.

Nei, M. and Roychoudhury, A.K. (1982) Genetic relationship and the evolution of human races. *Evolutionary Biology* **14**, 1–59.

Olson, T. (1978) Hominid phylogenetics and the existence of *Homo* in member 1 of the Swartkrans formation, South Africa. *J. Hum. Evol.* **7**, 159–178.

Olson, T. (1981) Basicranial morphology of the extant hominoids and Pliocene hominids. In: *Aspects of Human Evolution*, Stringer, C.B. (ed.), Taylor and Francis, London, pp. 99–128.

Olson, T. (1985) Cranial morphology and systematics of the Hadar Formation Hominids and *'Australopithecus' africanus*. In: *Ancestors: the Hard Evidence*, Delson, E. (ed.), A.R. Liss, New York, pp. 102–119.

Oxnard, C.E. (1984) *The Order of Man: a biomathematical anatomy of the Primates.* Yale University Press, New Haven and London.

Oxnard, C.E. (1987) *Fossils, teeth and sex: new perspectives on human evolution.* University of Washington Press, Seattle and London.

Parkes, P.A. (1986) *Current Scientific Techniques in Archaeology.* St. Martin's Press, New York.

Passingham, R.E. (1975) Changes in the size and organisation of the brain in man and his ancestors. *Brain Behav. Evol.* **11**, 73–90.

Passingham, R.E. (1982) *The Human Primate.* Freeman, Oxford and San Francisco.

Patterson, C. (1982) Cladistics and classification. *New Scientist* 303–306.

Peters, C.R. and McGuire, B. (1981) Wild plant foods of the Makapansgat area: modern ecosystems analogue for *Australopithecus africanus* adaptations. *J. Hum. Evol.* **10**, 565–583.

Peters, C.R. and O'Brien, E.M. (1981) The early hominid plant food niche: insights from an analysis of human chimpanzee and baboon plant exploitation in eastern and southern Africa. *Curr. Anthrop.* **22**, 127–140.

Pickford, M. (1985a) A new look at *Kenyapithecus* based on recent discoveries in western Kenya. *J. Hum. Evol.* **14**, 113–114.

Pickford, M. (1985b) *Kenyapithecus*: A review of its status based on newly discovered fossils from Kenya. In: *Hominid Evolution Past, Present and Future*, Tobias, P.V. (ed.), A.R. Liss, New York, pp. 107–114.

Pickford, M. (1986) *Kenyapithecus*: a review of its status based on newly discovered fossils from Kenya. In: *Hominid Evolution: Past, Present and Future*, Tobias, P.V. (ed.), A.R. Liss, New York, pp. 107–114.

Pilbeam, D.R. (1982) New hominoid skull material from the Miocene of Pakistan. *Nature* **295**, 232–234.

Pilbeam, D.R. and Gould, S.J. (1974) Size and scaling in human evolution. *Science* **186**, 892–901.

Pope, G.G. (1988) Recent advances in Far Eastern paleoanthropology. *Ann. Rev. Anthropol.* **17**, 43–77.

Pope, G.G. and Cronin, J.E. (1984) The Asian Hominidae. *J. Hum. Evol.* **13**, 377–396.

Potts, R. (1984) Hominid hunters? Problems of identifying the earliest hunter–gatherers. In: *Hominid Evolution and Community Ecology*, Foley, R. (ed.), Academic Press, London, pp. 129–166.

Prentice, M.L. and Denton, G.H. (1988) The deep sea oxygen isotope record, the global ice sheet system and hominid evolution. In: *Evolutionary History of the 'Robust' Australopithecines*, Grine, F.E. (ed.), Aldine de Gruyter, New York, pp. 383–404.

Rak, Y. (1983) *The Australopithecine Face*. Academic Press, New York.

Rak, Y. (1986) The Neanderthal: a new look at an old face. *J. Hum. Evol.* **15**, 151–164.

Rak, Y. (1990) On the differences between two pelvises of Mousterian context from the Qafzeh and Kebara Caves, Israel. *Am. J. Phys. Anthrop.* **81**, 323–332.

Rak, Y. and Arensburg, B. (1987) Kebara 2 Neanderthal pelvis: first look at a complete inlet. *Am. J. Phys. Anthrop.* **73**, 227–231.

Rightmire, G.P. (1979a) Implications of the Border Cave skeletal remains for later Pleistocene human evolution. *Curr. Anthropol.* **20**, 23–35.

Rightmire, G.P. (1979b) Cranial remains of *Homo erectus* from Beds II and IV Olduvai Gorge, Tanzania. *Am. J. Phys. Anthrop.* **51**, 99–115.

Rightmire, G.P. (1981) More on the study of the Border Cave remains. *Curr. Anthropol.* **22**, 199–200.

Rightmire, G.P. (1984) Comparison of *Homo erectus* from Africa and S.E. Asia. *Cours. Forschungs. Senckenburg* **69**, 83–98.

Rightmire, G.P. (1985) The tempo of change in the evolution of mid-Pleistocene *Homo*. In: *Ancestors: the Hard Evidence*, Delson, E. (ed.), A.R. Liss, New York, pp. 255–264.

Rightmire, G.P. (1986) Stasis in *Homo erectus* defended. Paleobiology **12**, 324–325.

Rightmire, G.P. (1990) *The Evolution of Homo erectus. Comparative anatomical studies of an extinct human species*. Cambridge University Press.

Robinson, J.T. (1953) Meganthropus, Australopithecus and hominids. *Am. J. Phys. Anthrop.* **11**, 1–38.

Robinson, J.T. (1956) The dentition of the Australopithecine. *Transv. Mus. Men.* **9**, 1–179.

Robinson, J.T. (1968) The origin and adaptive radiation of the Australopithecines. In: *Evolution und Homanisation* (2nd edn), Kurth, G. (ed.), Fischer, Stuttgart, pp. 150–175.

Robinson, J.T. (1972) *Early hominid posture and locomotion*. Chicago University Press.

Rodman, P.S. and McHenry, H.M. (1980) Bioenergetics and the origin of hominid bipedalism. *Am. J. Phys. Anthrop.* **52**, 103–106.

Rose, M.D. (1984a) A hominine hip bone, KNM ER 3228 from East Lake Turkana, Kenya. *Am. J. Phys. Anthrop.* **63**, 371–378.

Rose, M.D. (1984b) Food acquisition and the evolution of positional behaviour: the case of bipedalism. In: *Food Acquisition and Processing in Primates*, Chivers, D.J., Wood, B.A. and Bilsborough, A. (eds.), Plenum, New York, pp. 509–524.

Santá Luca, A.P. (1980) The Ngandong fossil hominids. *Yale Univ. Publ. Anthrop.* **78**, 1–175.

Schick, K.D. (1987) Modelling the formation of Early Stone Age artifact concentrations. *J. Hum. Evol.* **16**, 789–807.

Seigal, M.I. and Carlisle, R.C. (1974) Some problems in the interpretation of Neanderthal speech capabilities: a reply to Lieberman. *Am. Anthrop.* **76**, 319–323.

Senut, B. and Tardieu, C. (1985) Functional aspects of Plio-Pleistocene hominid limb bones: implications for taxonomy and phylogeny. In: *Ancestors: the Hard Evidence*, Delson, E. (ed.), A.R. Liss, New York, pp. 193–201.

Shipman, P. (1983) Early hominid lifestyle: hunting and gathering or foraging and scavenging? In: *Animals and Archaeological Hunters and their Prey*, Clutton-Brock, J. and Grigson, C. (eds.), BAR Int. Series 163. Oxford, pp. 31–40.

Shipman, P. and Harris, J. (1988) Habitat preference and paleoecology of *Australopithecus boisei* in Eastern Africa. In: *Evolutionary History of the 'Robust' Australopithecines*, Grine, F.E. (ed.), Aldine de Gruyter, New York, pp. 343–382.

Sibley, C.G. and Ahlquist, J.E. (1984) The phylogeny of the hominoid primates as indicated by DNA–DNA hybridisation. *J. Mol. Evol.* **20**, 2–15.

Sillen, A. and Kavanagh, M. (1982) Strontium and paleodietary research: a review. *Yearbook of Phys. Anthrop.* **25**, 65–90.

Simpson, G.G. (1945) The principles of classification and a classification of mammals. *Bull. Amer. Mus. Nat. Hist.* **85**, 1–350.

Simpson, G.G. (1975) Recent advances in methods of phyletic inference. In: *Phylogeny of Primates: a multidisciplinary approach*, Luckett, W.P. and Szalay, F.S. (eds.), Plenum, New York and London, pp. 3–20.

Singer, R. and Wymer, J.J. (1982) *The Middle Stone Age at Klasies River Mouth in South Africa*. Chicago University Press.

Skelton, R.R., McHenry, H.M. and Drawhorn, G.M. (1986) Phylogentic analysis of early hominids. *Curr. Anthrop.* **27**, 21–43.

Smith, F.H. (1983) A behavioural interpretation of changes in craniofacial morphology across the archaic/modern *Homo sapiens* transition. In *The Mousterian Legacy*. Trinkaus, E. (ed.), *Brit. Archaeol. Report* **164**, 141–164.

Smith, F.H. (1984) Fossil hominids from the Upper Pleistocene of central Europe and the origins of modern Europeans. In: *The origins of modern humans: a world survey of the fossil evidence*, Smith, F.H. and Spencer, F. (eds.), Liss, A.R. Liss, New York, pp. 137–210.

Smith, F.H. (1989) Dental development as a measure of life history in primates. *Evolution* **43**, 683–688.

Smith, F.H., Falsetti, A.B. and Donnelly, S.M. (1989) Modern human origins. *Yearbook of Phys. Anthrop.* **32**, 35–68.

Sonneville-Bordes, D. de (1963) Upper Paleolithic cultures in western Europe. *Science* **142**, 347–355.

Spuhler, J.N. (1988) Evolution of Mitochondrial DNA in monkeys, apes and humans. *Yearbook of Phys. Anthrop.* **31**, 15–48.

Stanley, S.M. (1979) *Macroevolution: Pattern and Process*. W.H. Freeman, San Francisco.

Stanley, S.M. (1981) *The New Evolutionary Timetable*. Basic Books, New York.

Stern, J.T. and Sussman, R.L. (1983) The locomotor anatomy of *Australopithecus afarensis*. *Am. J. Phys. Anthrop.* **60**, 279–318.

Straus, L.G. (1989) Age of the modern Europeans. *Nature* **342**, 476–477.

Stringer, C.B. (1978) Some problems in Middle and Upper Pleistocene hominid relationships. In: *Recent Advances in Primatology. 3 Evolution*, Chivers, D.J. and Joysey, K.A. (eds.), Academic Press, London, pp. 395–418.

Stringer, C.B. (1981) The dating of European Middle Pleistocene hominids and the existence of *Homo erectus* in Europe. *Anthropologie (Brno)* **19**, 3–14.

Stringer, C.B. (1984) The definition of *Homo erectus* and the existence of the species in Africa and Europe. *Cours. Forsch. Senckenberg* **69**, 131–143.

Stringer, C.B. (1985) Middle Pleistocene hominid variability and the origin of late Pleistocene humans. In: *Ancestors: the Hard Evidence*, Delson, E. (ed.), A.R. Liss, New York, pp. 289–295.

Stringer, C.B. (1986) The credibility of *Homo habilis*. In: *Major Topics in Primate and Human Evolution*, Wood, B.A., Martin, L. and Andrews, P. (eds.), Cambridge University Press, pp. 266–294.

Stringer, C.B. (1987) A numerical cladistic analysis for the genus *Homo*. *J. Hum. Evol.* **16**, 135–146.

Stringer, C.B. (1990) The emergence of modern humans. *Sci. Am.* **263(6)**, 68–75.

Stringer, C.B. and Andrews, P. (1988) Genetic and fossil evidence for the origin of modern humans. *Science* **239**, 1263–1268.

Stringer, C.B. and Grun, R. (1991) Time for the last Neanderthals. *Nature* **351**, 701–2.

Stringer, C.B., Howell, F.C. and Melentis, J.K. (1979) The significance of the fossil hominid skull from Petralona, Greece. *J. Archaeol. Sci.* **6**, 235–253.

Stringer, C.B., Hublin, J.J. and Vandermeersch, B. (1984) The origin of anatomically modern humans in western Europe. In: *The origins of modern humans: A world survey of the fossil evidence*, Smith, F.H. and Spencer, F. (eds.), A.R. Liss, New York, pp. 51–136.

Stringer, C.B., Grun, R., Schwarcz, H.P. and Goldberg, P. (1989) ESR dates for the hominid burial site of Es Skhul in Israel. *Nature* **338**, 756–758.

Suzuki, H. and Takai, F. (1970) *The Amud man and his cave site*. University of Tokyo Press, Tokyo.

Szalay, F.S. and Delson, E. (1979) *Evolutionary History of the Primates*. Academic Press, New York.

Tague, R.G. and Lovejoy, C.O. (1986) The obstetrics of AL 288-1 (Lucy). *J. Hum. Evolution* **15**, 237–255.

Tattersall, I. (1986) Species recognition in human palaeontology. *J. Hum. Evol.* **15**, 165–175.

Tattersall, I. and Eldredge, N. (1977) Fact, theory and fantasy in human paleontology. *Am. Sci.* **65**, 204–211.

Thorne, A.G. (1981) The centre and the edge: the significance of Australian hominids to African palaeoanthropology. In: *Proc. 8th Pan African Congress of Prehistory and Quaternary Studies*, Leakey, R.E.F. and Ogot, B.A. (eds.), TILLMIAP, Nairobi, pp. 180–181.

Thorne, A.G. and Macumber, P.G. (1972) Discoveries of Late Pleistocene Man at Kow Swamp, Australia. *Nature* **238**, 316–319.

Thorne, A.G. and Wolpoff, M.H. (1981) Regional continuity in Australian Pleistocene hominid evolution. *Am. J. Phys. Anthrop.* **55**, 337–349.

Tobias, P.V. (1967) *Olduvai Gorge Vol. 2: The Cranium and maxillary dentition of Australopithecus (Zinjanthropus) boisei*. Cambridge University Press.

Tobias, P.V. (1971) *The Brain in Hominid Evolution*. Columbia University Press, New York and London.

Tobias, P.V. (1978) The earliest Transvaal members of the genus *Homo* with another look at some problems of hominid taxonomy and systematics. *Zeitschrift für Morphologie und Anthropologie* **69**, 225–265.

Tobias, P.V. (1981) '*Australopithecus afarensis*' and *A. africanus*: a critique and an alternative hypothesis. *Palaeont. Africana* **23**, 1–17.

Tobias, P.V. (1987) The brain of *Homo habilis*: A new level of organisation in cerebral evolution. *J. Hum. Evol.* **16**, 741–761.

Tobias, P.V. (1991) *Olduvai Gorge Vol. 4* The skulls, endocasts and teeth of *Homo habilis*. Cambridge University Press.

Tobias, P.V. and von Koenigswald, G.H.R. (1964) A comparison between the Olduvai hominines and those of Java and some implications for hominid phylogeny. *Nature* **204**, 515–518.

Toth, N. (1985) The Oldowan Reassessed: A close look at early stone artifacts. *J. Archaeol. Sci.* **12**, 101–120.

Toth, N. (1987a) The First Technology. *Sci. Am.* **255**, 104–113.

Toth, N. (1987b) Behavioural inferences from early stone artifact assemblages: an experimental model. *J. Hum. Evol.* **16**, 763–787.

Trinkaus, E. (1983a) *The Shanidar Neanderthals*. Academic Press, New York.

Trinkaus, E. (1983b) Neanderthal postcrania and the adaptive shift to modern humans. In: *The Mousterian legacy*, Trinkaus, E. (ed.), *B.A.R. International Series 164*, pp. 165–200.

Trinkaus, E. (1984a) Neanderthal pubic morphology and gestation length. *Curr. Anthropol.* **25**, 509–551.

Trinkaus, E. (1984b) Western Asia. In: *The origins of modern Humans: a world survey of the fossil evidence*, Smith, F.H. and Spencer, F. (eds.), A.R. Liss, New York, pp. 257–293.

Trinkaus, E. (1987) The Neanderthal face: evolutionary and functional perspectives on a recent hominid face. *J. Hum. Evol.* **16**, 429–433.

Trinkaus, E. (1990) Cladistics and the hominid fossil record. *Am. J. Phys. Anthrop.* **83**, 1–12.

Trinkaus, E. and Smith, F.H. (1985) The fate of the Neanderthals. In: *Ancestors: the Hard Evidence*. Delson, E. (ed.), A.R. Liss, New York, pp. 325–333.

Turner, A. and Chamberlain, A. (1989) Speciation, morphological change and the status of African *Homo erectus*. *J. Hum. Evol.* **18**, 115–130.

Valladas, H., Geneste, J.M., Joron, J.-L. and Chadelle, J.P. (1986) Thermoluminescene dating of Le Moustier (Dordogne, France). *Nature* **322**, 452–454.

Valladas, H., Reybs, J.L., Joron, J.-L., Valladas, G., Bar-Yosef, O. and Vandermeersch, B. (1988) Thermoluminescene dating of Mousterian 'proto cro-magnon' remains from Israel and the origin of modern man. *Nature* **331**, 614–616.

Van Noten, F.L. (1983) News from Kenya. *Antiquity* 139–140.

Van Valen, L. (1986) Speciation and our own species. *Nature* **322**, 412.

Van Vark, G.N., Bilsborough, A. and Dijkema, J. (1989) A further study of the morphological affinities of the Border Cave I cranium, with special reference to the origin of modern man. *Anthropologie et Prehistoire* **100**, 43–56.

Vandermeersch, B. (1981) *Les hommes fossiles de Qafzeh, Israel*. CNRS, Paris.

Vandermeersch, B. (1989) The evolution of modern humans: recent evidence from southwest Asia. In: *The Human Revolution behavioural and biological perspectives on the origin of modern humans*, Mellars, P. and Stringer, C. (eds.), Edinburgh University Press, pp. 155–164.

Vrba, E. (1974) Chronological and ecological implications of the fossil Bovidae at the Sterkfontein australopithecine site. *Nature* **250**, 19–23.

Vrba, E. (1979) A new study of the scapula of *Australopithecus africanus* from Sterkfontein. *Am. J. Phys. Anthrop.* **51**, 117–129.

Vrba, E (1982) Biostratigraphy and chronology, based particularly on Bovidae of southern African hominid associated assemblages. *Pretirange; 1er Congrese Internat. Paleontol. Hum. Nice*, CNRS. pp. 707–752.

Vrba, E. (1988) Late Pliocene climatic events and hominid evolution. In: *Evolutionary History of the 'Robust' Australopithecines*, Grine, F.E. (ed.), Aldine de Gruyter, New York, pp. 405–426.

Wainscoat, J., Hill, A.V.S., Boyce, A.L., Flint, J., Hernandez, M., Thein, S.L., Old, J.M., Lynch, J.R., Falusi, A.G., Weatherall, D.J. and Clegg, J.B. (1986) Evolutionary relationships of human populations from an analysis of nuclear DNA polymorphism. *Nature* **319**, 491–493.

Walker, A. (1976) Remains attributable to *Australopithecus* in the East Rudolf succession. In: *Earliest Man and Environments in the Lake Rudolf Basin*, Coppens, Y., Howell, F.C., Isaac, G.L. and Leakey, R.E.F. (eds.), Chicago University Press, pp. 484–489.

Walker, A. (1981) Dietary hypotheses and human evolution. *Phil. Trans. Roy. Soc. B.* **292**, 57–64.

Walker, A. and Leakey, R.E.F. (1978) The hominids of East Turkana. *Sci. Am.* **239**, 54–66.

Walker, A., Leakey, R.E.F., Harris, J.M. and Brown, F.H. (1986) 2.5-myr *Australopithecus boisei* from west of Lake Turkana. *Nature* **322**, 517–522.

Wallace, J. (1973) Tooth chipping in the australopithecines. *Nature* **244**, 117–118.

Wallace, J. (1975) Dietary adaptations of *Australopithecus* and early *Homo*. In: *Paleoanthropology, Morphology and Paleontology*, Tuttle, R. (ed.), Mouton, The Hague, pp. 203–223.

Wallace, J. (1978) Evolutionary trends in the early hominid dentition. In: *Early Hominids of Africa*, Jolly, C.J. (ed.), Duckworth, London, pp. 285–310.

Walter, R.C. and Aranson, J.L. (1982) Revision of K/Ar ages for the Hadar hominid site, Ethiopia. *Nature* **296**, 122–127.

Ward, S. and Kimbel, W. (1983) Subnasal alveolar morphology and the systematic position of *Sivapithecus*. *Am. J. Phys. Anthrop,* **61**, 157–171.

Ward, S.C. and Pilbeam, D.R. (1983) Maxillofacial morphology of Miocene hominoids from Africa and Indo-Pakistan. In: *New Interpretations of Ape and Human Ancestry*, Ciochon, R.L. and Corruccini, R. (eds.), Plenum, New York, pp. 211–238.

Weidenreich, F. (1945) Giant early man from Java and South China. *Anthrop. Pap. Am. Mus. Nat. Hist.* **40**, 1–184.

Weidenreich, F. (1947) The trend of human evolution. *Evolution* **I**, 221–236.

Weidenreich, F. (1951) Morphology of Solo Man. *Anthrop. Pap. Am. Mus. Nat. Hist.* **43**, 205–290.

White, T.D. (1977) New fossil hominids from Laetoli, Tanzania. *Am. J. Phys. Anthrop.* **46**, 197–229.

White, T.D. (1980) Additional fossil hominids from Laetoli, Tanzania: 1976–1979 specimen. *Am. J. Phys. Anthrop.* **53**, 487–504.

White, T.D. (1984) Pliocene hominids from the Middle Awash, Ethiopia. *Cour. Forsch. Inst. Senckenberg* **69**, 57–68.

White, T.D. (1985) The hominids of Hadar and Laetoli: an element-by-element comparison of the dental samples. In: *Ancestors: the Hard Evidence*, Delson, E. (ed.), A.R. Liss, New York, pp. 138–152.

White, T.D. (1988) The comparative biology of 'robust' *Australopithecus*: clues from context. In: *Evolutionary History of the 'Robust' Australopithecines*, Grine, F.E. (ed.), Aldine de Gruyter, New York, pp. 449–484.

White, T.D. and Suwa, G. (1987) Hominid footprints at Laetoli: facts and interpretations. *Am. J. Phys. Anthrop.* **72**, 485–514.

White, T.D., Johanson, D.C. and Kimbel, W. (1981) *Australopithecus africanus*: its phyletic position reconsidered. *S. Afr. J. Sci.* **77**, 445–470.

Wind, J. (1978) Fossil evidence for primate vocalisations? In: *Recent Advances in Primatology 3: Evolution*, Chivers, D.J. and Joysey, K.A. (eds.), Cambridge University Press, pp. 87–92.

Wolpoff, M.H. (1980) *Paleoanthropology*. A. Knopf, New York.

Wolpoff, M.M. (1984) Evolution in *Homo erectus*: the question of stasis. *Palaeobiology* **10**, 389–406.

Wolpoff, M.H. (1985) Human evolution at the peripheries: the pattern at the eastern edge. In: *Hominid Evolution Past, Present and Future*, Tobias, P.V. (ed.), A.R. Liss, New York, pp. 355–366.

Wolpoff, M.H. (1986) Stasis in the interpretation of evolution in *Homo erectus*: a reply to Rightmire. *Paleobiology* **12(3)**, 325–328.

Wolpoff, M.H. (1988) Divergence between early hominid lineages: the roles of competition and culture. In: *Evolutionary history of the 'Robust' Australopithecines*, Grine, F.E. (ed.), Aldine de Gruyter, New York, pp. 485–497.

Wolpoff, M.H. (1969) Multiregional evolution: the fossil alternative to Eden. In: *The Human Revolution, behavioural and biological perspectives on the origins of modern humans*, Mellars, P. and Stringer, C. (eds.), Edinburgh University Press, pp. 62–108.

Wolpoff, M.H., Zinzhi, Wu and Thorne, A. (1984) Modern *Homo sapiens* origins: a general theory of hominid evolution involving the fossil evidence from East Asia. In: *The Origins of Modern Humans: A World Survey of the Fossil Evidence*, Smith, F.H. and Spencer, F. (eds.), A.R. Liss, New York, pp. 411–483.

Wolpoff, M.H., Spuhler, J.N., Smith, F.H., Radovcic, J., Pope, G., Frayer, D.W., Eckhardt, R.R. and Clark, G. (1988) Modern Human Origins. *Science* **241**, 772–773.

Wood, B.A. (1976) Remains attributable to *Homo* in the East Rudolf succession. In: *Earliest Man and Environments in the Lake Rudolf basin: Stratigraphy, Paleoecology and Evolution*, Coppens, Y., Howell, F.C., Isaac, G.L. and Leakey, R.E.F. (eds.), Chicago University Press, pp. 490–506.

Wood, B.A. (1978) Classification and phylogeny of East African hominids. In: *Recent Advances in Primatology Vol. 3. Evolution*, Chivers, D.J. and Joysey, K.A. (eds.), Academic Press, London, pp. 351–372.

Wood, B.A. (1984) The origin of *Homo erectus. Coursch. Forschungs. Senckenberg* **69**, 99–111.
Wood, B.A. (1985) Early Homo in Kenya and its systematic relationships. In: *Ancestors: the Hard Evidence*, Delson, E. (ed.), A.R. Liss, New York, pp. 206–214.
Wood, B.A. (1988) Are 'robust' Australopithecines a monophyletic group? In: *Evolutionary History of the 'Robust' Australopithecines*, Grine, F.E. (ed.), Aldine de Gruyter, New York, pp. 269–284.
Wood, B.A. (1991) *Koobi Fora Research Project Vol. 4: Hominid Cranial Remains.* Clarendon Press, Oxford.
Wood, B.A. (1992a) Origin and early evolution of the genus *Homo. Nature* (in press).
Wood, B.A. (1992b) Hominid species and speciation. *J. Hum. Evol.* (in press).
Wood, B.A. (1992c) Early *Homo*: how many species? In: *Species, species concepts and primate evolution.* Martin, L. and Kimbel, W.H. (eds.), Plenum, New York.
Wood, B.A. and Chamberlain, A. (1986) *Australopithecus* : grade or clade? In: *Major topics in primate and human evolution*, Wood, B., Martin, L. and Andres, P. (eds.), Cambridge University Press, pp. 270–248.
Wood, B.A. and Van Noten, F.L. (1986) Preliminary observations on the BK 8518 mandible from Baringo, Kenya. *Am. J. Phys. Anthrop.* **69**, 117–127.
Wood, B.A., Li, Y. and Willoughby, C. (1991) Intraspecific variation and sexual dimorphism in cranial and dental variables among higher primates and their bearing on the hominid fossil record. *J. Anat.* **174**, 185–205.
Wright, R.V.S. (1972) Imitative learning of a flaked stone technology—the case of an orang utan. *Mankind* **8**, 296–306.
Wynn, T. (1985) Piaget, stone tools and the evolution of human intelligence. *World Archaeol.* **17(1)**, 32–43.
Wynn, T. and McGrew, W.C. (1989) An ape's view of the Oldowan. *Man* **24**, 383–398.
Zihlman, A.L. (1989) Common ancestors and uncommon apes. In: *Human Origins.* Dwant, J.C. (ed.), Clarendon Press, Oxford, pp. 81–105.
Zihlman, A.L. and Lowenstein, J.M. (1983) *Ramapithecus* and *Pan paniscus*: significance for human origins. In: *New Interpretations of Ape and Human Ancestry*, Ciochon, R.L. and Corruccini, R.S. (eds.), Plenum, New York, pp. 667–694.

Additional references

Baringa, M. (1992) 'African Eve' backers beat a retreat. *Science* **255**, 686–687.
Hedges, S.B., Kumar, S., Tamura, K. and Stoneking, M. (1992) Human origins and analysis of mitochondrial DNA sequences. *Science* **255**, 737–739.
Maddison, D.R. (1991) *Systematic Zoology* **40**, 355.
Templeton, A.R. (1992) Human origins and analysis of mitochondrial DNA sequences. *Science* **255**, 737.
Vigilant, L., Stoneking, M., Harpending, H., Hawkes, K. and Wilson, A.C. (1991) African populations and the evolution of human mitochondrial DNA *Science* **253**, 1503–1507.

Index

Abri Pataud 187, 197–198
Acheulean assemblages 175–181
Africa 57–59, 62–67, 69–114, 148–152,
  162–169, 173–181, 209–212, 214–218,
  225–230, 323–334
Afropithecus 58
Afrotarsius 57
AL 200-1A 82
AL 288 71, 81, 92–95, 97
AL 333-1 82
AL 400-1A 82
altruism and selection 7
America, human occupation of 212–213
amino acids
  and dating 25
  and hominoid evolution 19, 63, 64
  and recent human origins 213–218
Amphipithecus 57
Amud 187, 200–201
anagenesis 8, 9, 10, 131–133, 160–161,
  174–175, 215–219, 225, 229–233
anatomically modern humans see
  Homo sapiens sapiens
Anthropoidea 18–21
Anyathian industry
apes 18–21, 225–226
  African 19, 21, 54, 55–56, 58–59,
    63–66
  Asian 19–21, 54, 55–56, 62–63, 65
  Miocene 13, 58–66, 225–226
Arago 163, 170, 171, 191
^{40}Ar–^{39}Ar dating 23, 24, 116
Asia 57, 59–63, 153–162, 172–174,
  179–180, 200–207, 217–219, 230–232
Atapuerca 163
Aurignacian tools 193, 195–198,
  203–204
Australia, human occupation of 206–209,
  218
Australopithecus 68–108, 130, 225

adaptation 103–108
body size 98–99
craniofacial anatomy 72–73, 87–91
development 97–98
diet 105–108
encephalisation 138–141
environments 103–108
evolutionary relationships 99–103
history 68–71
postcrania 91–96
sites 69–71
species 71–87
tool using 106–107
Australopithecus aethiopicus 79, 87, see
  also Lomekwi, Omo, KNM WT 17000
Australopithecus afarensis 70–71, 81–84,
  87, 226–228
  body size 98, 99
  craniofacial anatomy 88–89
  dating 70, 81, 84, 86
  development 97
  dispute over 83
  encephalisation 138–141
  environments 104, 105
  evolutionary relationships 99–103
  morphology 81–84
  postcrania 91–96
  sites 70, 71, 84–86
Australopithecus africanus 69, 70,
  226–228
  body size 98–99
  craniofacial anatomy 88–90
  dating 70, 73
  development 97–98
  diet 106–107
  encephalisation 138–141
  environment 104–107
  evolutionary relationships 99–103
  morphology 73–76
  postcrania 91, 94, 96

Australopithecus africanus cont'd
  sites 70, 73–74
  tool using 106–107
Australopithecus boisei 70, 71, 109, 110, 117, 135, 136, 160
  body size 98, 99
  craniofacial anatomy 90–91
  development 97–98
  diet 107–108
  encephalisation 138–141
  environments 105
  evolutionary relationships 99–103
  morphology 78–80
  postcrania 91–97
  tool making 136
Australopithecus crassidens 227–228
Australopithecus robustus 69, 70, 109, 110, 122, 135, 160, 227–228
  body size 98–99
  craniofacial anatomy 87–91
  development 97–98
  diet 107–108
  encephalisation 138–141
  environments 103–106
  evolutionary relationships 99–103
  morphology 75–78
  postcrania 91–97
  sites 69, 70, 74, 75
autapomorphous characters in *H. erectus* 13–18, 160–162, 227, 230 see also derived characters

baboons 64–66
Baringo 163
BC 1–5 210–211
Beijing 153 see also *H. erectus*, Zhoukoudian, Asia
Belohdelie (Middle Awash) 70, 84, 85
Biache 163, 172, 191
Bilzingsleben 163, 173
biomechanics 28–29
  australopithecine cranio facial 87–91
biospecies 12, 234
bipedalism 12, 46–50, 56, 65, 67, 225–226
  early *Homo* 114, 119–121, 132–133
  H. erectus 150–154
  in *Australopithecus* 91–96
  Neanderthals 190–191
BK 67 167
'black skull' (Lomekwi) see KNM WT 17000
Bodo 163, 174

Bodvapithecus 60
Border Cave 205, 209–212
Boule, M. 190–191
brain 12, 29, 38–44
  and speech 42–44
  and tools 138–142, 180–181
  archaic *Homo sapiens* 165–173
  asymmetry 42–44
  Australopithecus 72–84, 97, 99, 103, 139–140
  early *Homo* 112–130, 138–142
  endocasts 44
  expansion 39, 44
  Homo erectus 149–159, 155–160, 180–181
  Neanderthals 189–190, 222
Brain, C.K. 106
Brno 187, 199
Broca's area 42–44, 72, 140 see also brain, Wernicke's area
Broken Hill see Kabwe
Broom, R. 69–70
Buda industry 179

carbon-14 dating 22–23, 195–197, 199–203, 206–207
catarrhines 54, 56
Cave of Heaths, Makapansgat 163, 166–168
Ceboidea 57–58
Cercopithecoidea 18–21, 59, 66, 229 see also Old World monkeys
Chatelperonian tools 193, 195–198, 203–204
Chenchiawo see Lantian
Chesowanja 69, 175
chewing activity 34–37
chimpanzee 18–21 see also *Pan*
chronospecies 8, 9, 12, 173–174, 232
Clactonian industry 179
clade 12, 59, 100–102, 160–162, 174–175, 232
cladistics 13–18, 230–232, 234
  and affinity 13–16
  and classification 18–21, 225, 228
  cladogram 14–15
cladogenesis 12
classification, biological 18–21
co-evolution 7
Cohuna 205, 208 see also Kow Swamp
Combe Capelle 187, 197–198
community ecology 7, 64–66, 228–230
convergent evolution 11

Coppens, Y. 71, 84
Cossack 205, 208
Cromagnon 187, 197–198

Dali 163, 172–173, 205–206
Dar es Soltan 205
Dart, R.A. 68–70
dating 21–27
  absolute 22–27
  relative 21, 27
  see also carbon-14, potassium–argon,
  ^{40}Ar–^{39}Ar, thermoluminescence,
  electron spin resonance, amino acids,
  palaeomagnetism, Pleistocene, Pliocene
Dendropithecus 58
dentition
  and development 32–34
  archaic H. sapiens 165–170
  Australopithecus 71–91, 97–98,
    99–101, 106–108
  early Homo 112–128, 131–134
  hominid 31–43, 225
  Homo erectus 150–162, 164–165
  Homo sapiens sapiens 200, 206,
    219–223
  morphology 31–33
  Neanderthals 189–190, 199–200,
    219–222
  structure 32–34
  Tertiary hominoid 59–62, 225
derived characters 11
Developed Oldowan 177–181 see also
  Oldowan, Acheulean
Die Kelders Cave 205
Djetis fauna, Java 154–158 see also
  Puchangan Beds
DNA 55, 63, 214–218, 233–234
  and modern human origins 214–215,
    217–218, 233–234
  mitochondrial 55
  nuclear 55, 63–64, 214
Dolni Vestonice 187, 199
Dryopithecus 59
  D. fontani 59
  D. laeitanus 59
Dubois, E. 154

early Homo see Homo, early
Eemian interglacial 184–185
electron spin resonance dating 24–25,
  195, 200–202, 211–212
Eliye Springs 163, 166–168, 209
encephalisation 38–40 see also brain

early Homo 138–140
H. erectus 180–181
in Australopithecus 138–140
Eocene primates 56–57
Equus Cave 205
euprimates 56
Europe 59–61, 63, 147, 163, 170–172,
  173, 178, 184–187, 191–200, 203–204,
  214–215, 220–223, 232–233
evolution 6
  evolutionary mechanisms 6–8
  evolutionary models 8–10
  mosaic 12, 232
evolutionary systematics 15–21
extensors
  of arm and hand 52, 53
  of thigh and leg 49–51, 91–92

face 12, 29–31, 34–37
  archaic H. sapiens 165–173
  Australopithecus 73–83, 87–91
  early Homo 112–124
  H. erectus 150–159
  H. sapiens sapiens 197–202,
    207–212, 219–223
  Neanderthals 189–190, 198–202,
    219–223
Fayum primates 57–58
femur 47–50 see also hind limb, leg
  Australopithecus 91–96
  early Homo 114, 121
  H. erectus 154, 160
flexors
  of arm and hand 52–53
  of thigh and leg 49–51, 91–92
Florisbad 163, 167–168, 205, 209
Fontéchevade 163, 171–172
foot 47–50
  arches 47, 48
  as lever 48
  as plate 47, 48
  Australopithecus 92–96
  early Homo 112–113
  Neanderthal 190
footprints, Laetoli 92, 93
forelimb 50–53
  and tool making 136
  ape 55
  Australopithecus 92–96
  early Homo 114, 121, 132
  Homo erectus 150–151
  Neanderthal 190–191
Fort Ternan 62–63

Gadeb   175–178
geladas   64–66
Gigantopithecus   62–63
  G. blackii   62
  G. giganteus   62
gluteus muscles   49–50
Gongwangling   see Lantian
Gorilla   18, 21, 54, 55, 225–227, 228
grades   13, 109, 130–131, 133, 145,
  149–152, 160–162, 233–234
Graecopithecus   see *Ouranopithecus*
Grenzbank (Java)   154–156
Grotte des Enfants   187, 197–198

Hadar   70, 71, 81, 83, 86, 91–95,
  104–105, 136, 143, 226–228
Hahnofersand   187, 199
hand   52, 53, 95–96, 112, 121, 190   see
  also forelimb, power grip, precision
  grip, thumb, tool making
Haplorhini   54, 57
Haua Fteah   205, 209
head, hominid   28, 29–44   see also
  skull, face, brain, vocalisation, dentition
Heidelberg   see Mauer
Hennig, W.   13
Hexian   163
hindlimb   46–50, 67, 91–96, 121,
  150–152, 154, 160, 190–191, 201–202
  see also bipedalism, leg, hip bone,
  pelvis
hip bone   46–50, 91–96, 121, 150–152,
  171, 190   see also hind limb, pelvis
holophyly   16   see also monophyly,
  paraphyly, polyphyly
Hominidae   56
  origins   64–67
hominids   7, 13, 18–21, 54, 56, 66   see
  also *Australopithecus, Homo*
  functional anatomy   28–53
  origins   64–67, 225–226
Hominoidea   18–21, 57–66, 229–230
Homo aethiopicus   129–130
Homo, early   109–144, 145, 227–228
  see also *H. habilis, H. ergaster,*
  H. erectus, H. rudolfensis
Homo erectus   119–166, 173–181,
  229–233
  sexual dimorphism   226–227
  stasis   232
Homo erectus, African   110–111,
  119–121, 122, 123–128, 131–134, 145,

148–152, 160–162, 228–231   see also
  H. ergaster
Homo erectus, Asian   153–162, 230
Homo ergaster   125–133, 148–149, 162,
  228–230   see also *H. erectus* (African)
Homo habilis   109, 110, 112–114,
  122–135, 140, 145, 154, 158, 160, 164,
  224, 228–229   see also *Homo,* early,
  H. rudolfensis, H. ergaster
Homo rudolfensis   128–134, 145, 152,
  160, 228–229   see also *Homo,* early,
  H. habilis, H. ergaster
Homo sapiens, archaic   113, 164,
  165–172, 174–175, 229–232
  Asian   172–175
  European   170–175
  North African   169, 173–175
  Sub-Saharan   167–168, 173–175
Homo sapiens rhodesiensis   167–168, 209
Homo sapiens sapiens   182–223
  alternative models for origin   215–219,
    233–234
  evolution within   222–223
  transition to   219–221
homoplasy   11, 228–229
Howell, F.C.   84
Hughes, A.R.   70, 123
humerus   51–52, 94–95
Hylobates   54
Hylobatidae   18–21, 54

Indonesia, *H. erectus*   153–158   see
  also *H. erectus*, Asian
Innominate   see hip bone
Isernia   179

Jacob, T.   154–155
Java   see Indonesia, *H. erectus*, Asian
jaw, hominid   29, 31   see also teeth
Jebel Irhoud   168–169, 205, 209
Jebel Qafzeh   187, 200–203, 217–218
Johanson, D.C.   71, 81

K.3   198
K.6   198
K.58   198
K.59   198
Kabuh Beds, Java   154–158   see also
  Sangiran
Kabwe   163, 166–168
Kanapoi   69, 70, 71, 84, 85
Kebara (Mt Carmel)   187, 200–202,
  221–222

Kebibit 163, 169
Kenyapithecus 62–63
Kilombe 175–181
Klasie's River Mouth 205, 209–212, 216
knee 47–50, 91, 94–95
KNM ER 406 80
KNM ER 730 117, 119, 162
KNM ER 732 79
KNM ER 992 117, 119, 125, 162
KNM ER 1470 117, 118, 125–132
KNM ER 1500 121
KNM ER 1590 117, 118, 125
KNM ER 1801 117, 118–119, 125
KNM ER 1802 117, 118–119, 125
KNM ER 1805 117, 118, 125–132
KNM ER 1813 117, 125–132
KNM ER 3228 120–121, 132, 160
KNM ER 3732 117, 118, 125–132
KNM ER 3733 117, 120, 121, 125, 160, 161, 162
KNM ER 3735 121, 134
KNM ER 3883 117, 120, 121, 125, 162
KNM WT 15000 119, 148, 150–152, 228–230 see also Nariokotome, *H. erectus*, African
KNM WT 17000 79, 80, 87, 90, 100, 103, 227–228
knuckle-walking 55, 59, 64, 96, 225
Koenigswald, G.H.R. von 154
Koobi Fora 69, 70, 86, 105, 114–121, 124–135, 138, 141–142, 161–162
Kow Swamp 205, 207–208
Kow Swamp 1 207
Kow Swamp 5 207
Krapina 187, 198–199
KRM 13400 210
KRM 16424 210
KRM 21776 210
KRM 41815 210
Kromdraai 69, 70, 75, 78, 227
Kulna 187, 199

L 894-1 121
La Chaise 163, 172, 191
La Chapelle aux Saints 187, 190–191, 195–196
La Ferrassie 187, 191, 195–196
La Quina 187, 191, 195–196
Laetoli 69, 70, 71, 81, 86, 92–93, 104, 163, 226–228
Lake Mungo 205, 207–209 see also Mungo I–III

Lake Rudolf see Lake Turkana
Lake Turkana (Rudolf) 114–121, 136, 228 see also Koobi Fora, Lomekwi, Nariokotome
language 37–38, 42–44, 72, 138–140, 221–222
Lantian 158–159, 179
Lazaret 163, 172
Leakey, L.S.B. 70, 111–114
Leakey, M.D. 71, 111–114
Leakey, R.E.F. 84, 85, 115–120
Le Moustier 187, 195
LH 4 82
Liujiang 205–206
Lomekwi (West Turkana) 70, 79, 100–103 see also KNM WT 17000
Lothagam 69, 70, 71, 84, 85
Lower Pleistocene 229–230
'Lucy' see AL 288
Lufeng 61, 62, 226–227
Lufengpithecus 61, 62
Lukeino 66

Maba 163, 173, 205–206
Maka (Middle Awash) 70, 84, 85
Makapansgat 69, 70, 73, 74, 104, 106–107
mandible, hominid 29, 31, 34–37
masticatory apparatus 34–37, 87–91, 225 see also skull, dentition
Mauer 163, 170–171, 173
maxilla, hominoid 29, 30, 36, 37, 62–63
Meganthropus palaeojavanicus 157
Melka Kunturé 175–178
Middle Awash 69, 70, 86, 175–178
Middle Palaeolithic assemblages 183, 191–195, 203–204, 222–223
Middle Pleistocene 191, 230–232, 234
Miocene
  environments 58–62, 65–66
  primates 58–67, 229–230
Mladec 187, 199
Modjokerto 153, 155
molecular evolution 19, 63, 64, 233–234
monkeys 7, 18, 21, 64–66 see also Cercopithecoidea, Ceboidea
monophyly 15–21 see also holophyly, paraphyly, polyphyly
Montmaurin 163, 164, 172, 191
morphospecies 12, 234
Mousterian industries 192–204
Mt Carmel see Kabara, Skhul, Tabun
Mt Circeo 187, 195

Mungo see also Lake Mungo
  I 207–208
  III 207–289

Nariokotome, West Turkana 115,
  119–121, 148, 150–152, 180, 228–230
  see also KNM WT 15000, *H. erectus*,
  African
Narmada 163, 172, 206
Ndutu 163, 168, 173
Neanderthals 186–204, 217–218,
  233–234
  central and eastern Europe 198–200
  contexts 191–195
  development 222
  Middle East 200–204
  morphology 189–191
  selection and modern morphology
    219–222
  vocalisation 221–222
  western Europe 195–198
Ng I 156
Ng 6 156
Ng 12 156
Ngaloba 163, 166–168, 209
Ngandong 155–158, 174, 206
neurocranium, hominoid 29–31
New World monkeys see Ceboidea
Niah 205–206, 209
Notopuro Beds, Java 154–158

OH 5 80, 112
OH 7 111–114
OH 8 112, 113
OH 13 112, 113, 124, 164
OH 16 114, 124
OH 24 113, 114
OH 28 164
OH 48 112
OH 62 114, 121, 134, 228
Oldowan industries 138–139, 142–143,
  175–181
  developed Oldowan 177–181
Olduvai Gorge 69, 110–114, 123–124,
  125, 128–136, 138–139, 141, 175–178
Old World monkeys see
  Cercopithecoidea
Oligocene 57
Oligopithecus 57
Olorgesailie 175–181
Omo 69, 70, 71, 84–87, 121, 137, 205,
  209
Omo (Kibish) 205, 209–212

Omo (Kibish) 1 205, 209–211
Omo (Kibish) 2 168, 205, 209–211
orang see *Pongo*
Oreopithecus
Ouranopithecus 60, 61, 63, 225

Pajitanian industry 180
palaeomagnetism 25–27, 73–74, 148
Pan 54, 55, 63–66, 96, 225
P. paniscus 55, 96
P. troglodytes 55
parallel evolution 11
  in *Australopithecus* 100–103,
    227–228
  in *H. erectus* and *H. sapiens* 174–175
paraphyly 15–21, 56 see also
  holophyly, monophyly, polyphyly
Parapithecidae 57
paromomyids 56
Pavlov 187, 199
pelvis 46–50
  in *Australopithecus* 91–97
  H. erectus 150–152, 180–181
  Neanderthal 190, 222
Perning see Modjokerto
Petralona 163, 164, 170–171
phenetic resemblance 11, 17, 231
phenotype, lability of, in evolution of
  modern humans 221–223
phyletic gradualism 8, 9, 10, 225,
  231–232
phyletic principles 10–18
  inference 11–18
  weighting 11–18
Pithecanthropus 154–158 see also
  H. erectus, Asian
plesiadapids 56
Plesiadapiformes 56
plesiomorphous characters 13–18, 62,
  227–228, 230–232, see also primitive
Pliocene
  environments 64–66, 103–108
  hominids 68–108, 233
  hominoids 64–67
  monkeys 64–66
  see also *Australopithecus*, early *Homo*,
    Koobi Fora, Olduvai, Omo
Pliopithecus 59
polyphyly 16 see also holophyly,
  monophyly, paraphyly
Pondaungia 57
Pongidae 18–21, 54, 55 see also apes,
  Gorilla, Pan, Pongo

Pongo 18–21, 54–55, 62–65, 227
potassium–argon dating 23–24, 71, 81, 110–111, 116–117, 148–152
power grip 52, 53
precision grip 53 see also hand, tool-making
Predmost 187, 197, 199
Primates 18–21, 56–65
primitive characters 11
Proconsul 58–59
  P. africanus 58, 61
  P. major 58
Proconsulidae 58–64, 65
Propliopithecidae 57
Prosimii 18–21
Puchangan Beds, Java 154–158 see also Sangiran
punctuated equilibria 8–10, 162, 164–165, 224, 231–232

Qafzeh 6 202
Qafzeh 9 202 see also Jebel Qafzeh

Ramapithecus 60, 62, 225
regional continuity model 215–219
reticulate evolution 215–219, 224
Rift Valley, East Africa 64–66, 69, 71, 115–116, 137–138, 177–178
rotary chewing 67
Rudapithecus 60

sacrum 46–47, 121
Saldanha 163, 167–168
Salé 163, 168, 169, 173
Sambungmachan 155, 156, 158
Samburu Hills 66
Sangiran 153–158
Sangiran 1 153
Sangiran 4 153, 156
Sangiran 6 156
Sangiran 17 153
Sartono, S. 155
scapula 51–52
selection, group 7
selection, kin 7
selection, natural 6, 7, 10
  and modern human origins 215–223
  K-type 7
  r-type 7
Senga 5a 138
sexual dimorphism
  in *Australopithecus* 226–228
  in Miocene hominoids 225

patterns 227
Shanidar 187, 200–202
Shanidar 1 201
Shanidar 2 201
Shanidar 5 201
shoulder 50–52 see also forelimb 94–96, 121, 151–152, 190
Sidi Abderrahman 163, 179
Simiolius 58
Singa 205
single origin model 215–219
Sivapithecus 60, 61–63, 225
Sivapithecus darwini 60
Sivapithecus indicus 60, 61
Sivapithecus meatei 60–61
Sivapithecus sivalensis 60
Siwaliks 60–62
SK 12 77
SK 15 164
SK 23 77
SK 46 77
SK 48 77
SK 80 96
SK 847 114, 121–123
SK 876 77
SK 3155 96
Skhul (Mt Carmel) 187, 200–203, 217–281
Skhul 4 202
Skhul 5 202
Skhull, hominid 29–44
  archaic *Homo sapiens* 165–174
  Australopithecus 71–91, 100–103
  early *Homo* 112–130, 132–133
  Homo erectus 119–121, 149–165
  Homo sapiens sapiens 197–202, 204–212, 219–222
  Neanderthals 189–204, 219–222
socioecology, early hominid 142–144, 228–229
Solo River 154–158 see also Ngandong, Trinil
speciation 7, 8, 10, 12, 234
  allopatric 8, 10
  and modern human origins 215–223, 224
  in *Australopithecus* 100–103
  in early *Homo* 132–134, 160–162
  rates 8
  selection 8, 10
  sympatric 8
species recognition, hominid 233
  in *Australopithecus* 69–87, 100–103

species recognition, hominid *cont'd*
in early *Homo* 112–130, 132–134,
149–152, 160–162
Spy 187, 191
St Cesaire 187, 195–196
stasis, in *H. erectus* 158, 162, 164–165,
232
Steinheim 101, 163, 170–172
Sterkfontein 69, 70, 73, 74, 78, 104,
122–123, 226–228
stratophenetic approach 11–14, 225
Strepsirhini 54, 56
Sts 5 76
Sts 7 95
Sts 14 95, 97
Sts 36 76
Sts 52a,b 76
Stw 53 114, 122–123
Sts 71 76
Stw 252 75, 96, 252
Swanscombe 163, 170–172, 191
Swartkrans 69, 70, 74, 75, 78, 104–106,
121–122, 135, 227
Symphalangus 54
synapomorphous characters 13–18 see
also derived characters
Szelethian industry 203

Tabarin 70, 84, 85
Tabun (Mt Carmel) 187, 200–202
Taieb, M. 71
taphonomy 2, 29, 74, 104–106,
141–142, 180, 191, 193
Tarsius 54, 57
Taung 69, 70, 73, 104
taxonomy 18–21, 230
Tayacian 179
teeth see dentition
Ternifine 164, 169, 179
thermoluminescence dating 24–25, 195,
200–202
Thomas Quarries 163, 169
thumb, hominid 52, 53
Tighenif see Ternifine
Tobias, P.V. 70, 123, 124
tool making 52–53, 136–144, 167,
175–181, 183, 192–195, 197–198,
203–204, 207, 229 see also forelimb,

hand, Oldowan, Olduvai, Hadar, Koobi
Fora, Omo, power grip, precision grip
Trinil 153–158
truncal erectness 12, 29, 67
trunk 67, 91–93, 119–121, 132–133,
150–152, 190–191
Tugan Hills 66
Turkanopithecus 58

Ubeidiya 179
Uluzzian industry 203
Upper Palaeolithic industries 183,
191–195, 203–204, 222–223 see also
Chatelperronian, Aurignacian,
Gravettian
Upper Pleistocene 182–223, 232–234
chronology and environment 184–186

Vallonet 179
Velika Pecina 187, 199
Vertesszollos 163, 170, 173
Victoriapithecus 59, 66
Vindija 187, 198–199, 204
vocalisation 37, 38, 42–44, 138–140,
180–181
Neanderthal 221–222

Wadjak 205–206
Weidenreich, F. 158–160, 217
Wernicke's area 42–44 see brain,
Broca's area
West Turkana see KNM WT 15000,
KNM WT 17000, Lake Turkana,
Lomekwi, Nariokotome
White, T.D. 81–84, 104–105, 114
Willandra Lakes 207–208 see also
Lake Mungo
Wurm glaciation 184–186, 190–204

Zhoukoudian 158–160, 179–180
Zhoukoudian (Upper Cave) 205–206
ZKD see Zhoukoudian
ZKD G1 159
ZKD H1 159
ZKD I 159
ZKD III 159
ZKD X 159
Zuttiyeh 187, 200–201